THE JOURNEYS OF TREES

ADDITIONAL PRAISE FOR *THE JOURNEYS OF TREES*

"A handful of seeds is all it takes to launch a new forest. Should we do it? Zach St. George chases the answer to this vital question from California to New Zealand in this smart and engaging book that illuminates the past and explores possible futures."

—Kim Todd, author of *Tinkering with Eden*

"In the rings of trees, we can see the past. Zach St. George brilliantly reports and narrates a journey to understand the future of forests trying to stand tall against the forces of climate change—and explores how we, a powerful hand holding a tiny seed, can help."

—Harley Rustad, author of *Big Lonely Doug*

"A beautifully written and deeply reported look at the world of trees at this moment of crisis. One of the best and most engaging pictures I've seen of how climate change has impacted and will impact the story of the species facing it."

—James Pogue, author of *Chosen Country*

THE JOURNEYS OF TREES

A STORY ABOUT FORESTS, PEOPLE, AND THE FUTURE

Zach St. George

Printed in the United States of America
First Edition

For information about special discounts for bulk purchases, please contact W. W. Norton Special Sales at specialsales@wwnorton.com or 800-233-4830

Manufacturing by LSC Communications, Harrisonburg
Book design by Chris Welch
Production manager: Lauren Abbate

Library of Congress Cataloging-in-Publication Data

Names: St George, Zach, author.
Title: The journeys of trees : a story about forests, people, and the future / Zach St. George.
Description: First edition. | New York, NY : W. W. Norton & Company, [2020] |
Includes bibliographical references and index.
Identifiers: LCCN 2020005951 | ISBN 9781324001607 (hardcover) | ISBN 9781324001614 (epub)
Subjects: LCSH: Forest conservation—United States.
Classification: LCC SD412 .S74 2020 | DDC 333.75/160973—dc23
LC record available at https://lccn.loc.gov/2020005951

W. W. Norton & Company, Inc., 500 Fifth Avenue, New York, N.Y. 10110
www.wwnorton.com

W. W. Norton & Company Ltd., 15 Carlisle Street, London W1D 3BS

1 2 3 4 5 6 7 8 9 0

To Janet and Judith, David and George

CONTENTS

THE JOURNEYS OF TREES

INTRODUCTION

In February 2004, Connie Barlow lay down under a small evergreen tree, stared up through its branches, and made it a promise. "I'm going to find a way," she told the tree. "I'm going to move you."

Barlow, a retired science writer and working evolutionary evangelist, was in a state park overlooking the Apalachicola River, which cuts north to south across the Florida Panhandle. The tree was a *Torreya taxifolia*, called variously in English "Florida nutmeg," "savine," "gopherwood," and "stinking cedar," the last for the offensive smell its foliage emits when crushed. Barlow, a partisan, thinks the nickname is what's offensive. She prefers to call the tree, simply, "Florida torreya."

The tree is a conifer, with pin-sharp needles and fleshy fruit. From afar, the tree has the draped, disheveled form of an old shag rug and is of medium build, sometimes fifty or sixty feet tall, although it's hard to find one that tall today. It is rare, growing over only a couple dozen square miles, in the steep ravines that cut through the limestone bluffs overlooking the Apalachicola River.

At the beginning of the twentieth century, the Florida torreya was confined to its few square miles, but within these miles it was common. Locals used torreyas for fenceposts and firewood and Christmas trees. But in the 1950s, the torreyas started dying. The culprit seemed to be a fungus. The trees died back to their roots, which then sent up suckers. These suckers would get to be ten or twelve feet tall and a couple inches thick and then they would succumb to the fungus, too. What was the fungus? Why now? Nobody knew.

Decades went by, and the tree grew rarer and rarer. Various people and universities and botanical gardens and state and federal agencies tried to save it, but none of them could wrest it from its terminal slide. Then, in the early 2000s, Connie Barlow visited Torreya State Park, found the suckers, and saw that the problem was a simple one—so simple, in fact, that it beggared belief that nobody had realized it and acted on it before. The problem wasn't the fungus at all. The tree's real problem was that it had gotten stuck.

Individual trees, as will be apparent to most people who have encountered one, don't often move around. Once a tree is rooted in one spot, it's rare for it to wind up in another. Forests, though, are restless things. Anytime a tree dies or sprouts, the forest it is a part of has shifted a little. The migration of a forest is just many trees sprouting in the same direction. Through the fossils that ancient forests left behind, scientists can track their movements over the eons. They shuffle back and forth across continents, sometimes following the same route more than once, like migrating birds or whales.

This Florida torreya, Barlow thought, was not supposed to be a Florida torreya at all. It should have been a northern Georgia torreya, or a Virginia torreya, or maybe even a Pennsylvania torreya. It should have migrated north at the end of the Pleistocene, 11,700 years ago, but for some reason, it had not. Now it was marooned in hot, sticky Florida, a place growing only hotter and stickier. The fungus was just a symptom of this mismatch, the way polar bears in zoos will sometimes grow algae on their fur and turn green. The Florida torreya, she thought, was a tree in the wrong place. The solution to this problem was simple: Move the tree north.

In his 1908 book *Worlds in the Making*, Swedish scientist Svante Arrhenius wrote that "the enormous combustion of coal by our industrial establishments suffices to increase the percentage of carbon dioxide in the air to a perceptible degree." He predicted this would cause the planet to warm. He saw it as a welcome side effect. "We may hope to enjoy ages with more equable and better climates," he wrote, "ages when the earth will bring forth much more abundant crops than at present, for the benefit of rapidly propagating mankind." By the 1980s, scientists thought Arrhenius was correct: Human pollution was warming the world. But it seemed likely to be less pleasant than the Swede had imagined. In a 1981 paper, a group of NASA scientists led by atmospheric physicist James Hansen predicted rising seas, growing deserts, and "large-scale human dislocations."

The rest of the world's living things would be rearranged, too. Every species has certain climatic requirements—what degree of heat or cold it can tolerate, for example. When the climate changes, the places that satisfy those requirements change, too. Species are forced to follow. All creatures are capable of some degree of movement—or dispersal, as scientists call it. A mobile creature, like a fish or a bird, might move around throughout its life. But even creatures that appear immobile, like trees and barnacles, are capable of dispersal at some stage of their life—as a seed, in the case of the tree, or as a larva, in the case of the barnacle. A creature must get from the place it is born—often occupied by its parent—to a place where it can survive, grow, and reproduce. Movement is constant and essential.

From fossils, scientists knew that even creatures like trees had moved with surprising speed during past periods of climate change. But the climate now seemed to be changing much faster than ever before. By the early 1990s, a growing number of scientists had concluded that many species would not be able to keep up.

They suggested the same solution that Connie Barlow later would:

Move the species to places where they would be better suited. People had spent millennia moving living things from one part of the world to another, farming and gardening and trying to improve their surroundings, all with little thought for the fate of the species involved. In one sense, then, helping slow species move faster or helping fast species avoid barriers would be only a small variation on this oft-repeated theme.

But many more scientists—often the same scientists who suggested it—were wary of this solution. The species that people had moved from one part of the world to another had often caused problems both for people and for other species. They often turned into invaders. Scientists and conservationists had argued for decades that people should avoid moving species from one place to another. They should aim to keep the world's living things in roughly their current arrangement—not static, but within some historic range of variation. The best way to help the world's living things cope with climate change, they said, would be to preserve wild places, to maintain or create corridors where species might migrate on their own, or best of all, to stop the world from warming further. Throughout the 1980s and '90s, scientists treated the idea of moving species to save them from climate change as a subject for more discussion, as a possibility for another day.

Barlow told me that for some forms of life, there was room for deliberation. "You've got time for perennials and salamanders and things you can just pick up and move," she said. But trees were slow. Creating a new forest demands a certain amount of foresight, a departure from the usual human schedule. Barlow saw her plan to move the Florida torreya north as a kind of optimism. Where others denied the depth of the problem, she accepted it. With acceptance came agency. She couldn't save the tree from the fungus. She couldn't keep the climate from changing. Really, there weren't very many things she *could* do. But she could plant a tree.

This is a book about trees, so it's worth noting here at the outset a point of linguistic sloppiness: The word "tree" is really only a description of geometry. Scientists will sometimes use a more exacting definition, but for our purposes the common one will do: A tree is a tall plant with a woody stem. In a way, a tree is just a logical conclusion; for an organism in the business of harvesting solar energy, it helps to have a platform higher than the neighbor's. Tall, woody plants have evolved many times in Earth's history. The oldest known trees are preserved as fossils in upstate New York. They date to the Middle Devonian, some 387 million years ago. These pseudosporochnalean trees seem to have looked something like enormous stalks of celery, long fluted columns topped with bunches of fingerlike photosynthetic appendages. The lycopsids were another group of ancient trees. Lycopsids today are club mosses, spike mosses, and quillworts, modest things, but the lycopsids of old grew a hundred feet tall and made globe-spanning forests. Some of these trees looked like they were made of tarantula legs. Others looked like golf tees that grew a thick coat of hair.

More familiar trees showed up in the Permian, 299 million years ago. This was likely the earliest point in the planet's history when a time-traveling botanist could set their machine down on land and feel nearly at home. They might write to their friends in the future and say, "You will begin to think that I manufacture Pines at my pleasure." Many of the trees they would have seen were conifers, ancestors of the trees that still cover much of the planet—the pines and larches, kauri and sequoias. The six hundred–some extant conifer species include the world's biggest, tallest, and oldest trees.

Over the next 100 million years, more members of the modern ensemble made their way onstage. One was the cycad. Often confused with palm trees, cycads have palmlike fronds and bristled trunks. Today they can frequently be found growing in corporate atriums. Another was the ginkgo, a tree with leaves like ducks' feet, currently represented by a single species, *Ginkgo biloba*. Monks in China and Japan tended the tree for centuries, planting it outside temples. Indeed, wrote British biologist Colin Tudge, it is unclear whether any wild trees remain or whether all the surviving specimens were planted by people. Ginkgoes are now popular as street trees everywhere they will grow.

These three types of trees are all gymnosperms. They cast sex to the wind, releasing clouds of pollen from male cones in the hope that they will find corresponding female cones or ovules. Gymnosperms ruled the world for 150 million years before angiosperms appeared. Angiosperms had a different strategy. Many of them relied on other creatures to join pollen and egg. They made lures, colorful and sugary and aromatic, called flowers. Over the ages, flowering trees usurped the gymnosperms. Today they far outnumber gymnosperms in their diversity of species, and they dominate forests across much of the world. It seems like an understatement to describe this grand competition as one between two simple categories of trees. As ecologist E. C. Pielou wrote, gymnosperms and angiosperms are deeply unalike. "The differences between them are profound," she wrote, "so profound, that it is ludicrous, nowadays, to lump them collectively as 'trees.'" But she offered no substitutes.

Tall and woody, then. Somewhere in that unparticular set of characteristics is something that particularly stirs us. Although a story of climatic mismatch could be told of any number of creatures—with a quillwort or a coral as star, for instance—it would bump endlessly into little hurdles of understanding. Trees are more familiar. They are everywhere, covering a third of the land on Earth. Forests are home to millions of species, so a story about trees says much about living things in general. And people like trees. They provide us fuel and building supplies, true, but the human-arboreal connection goes beyond the merely practical. We seem to have a stronger affinity for trees than we do for grasses, which supply most of our food, or for algae, which supply most of the world's oxygen. Maybe it is instinct. Some have speculated that a scene with the right mix of trees and open land summons some deep nostalgia for the savanna of our species' youth. Or maybe it's just a weird type of narcissism. As historian Jared Farmer noted, trees are identifiable as individuals and often take a humanlike shape, with limbs and a trunk. You can hug a tree. Try hugging a quillwort.

A tree is also a ready symbol, embodying different qualities at different times in its life. A seed, for example, suggests hidden potential: See the mighty oak spring from the tiny acorn. A sapling that bends in the wind, meanwhile, demonstrates a flexibility that people would do well to emulate. A mature tree, most evocative of all, represents endurance. "The most interesting ideas connected with trees are those suggested by their stability and duration," wrote botanist Asa Gray in the mid-1800s. "They far outlast all other living things, and form the familiar and appropriate symbols of long-protracted existence." He continued, quoting from the Book of Isaiah: "'As the days of a tree shall be the days of my people' is one of the most beautiful and striking figures under which a blessing can be conveyed."

Planting a tree, then, is a symbolic act. A tree might outlive the person who planted it by centuries or millennia, providing shade or beauty or raw materials to generations unborn. Hydroseed your yard and nobody will thank you. But plant a tree, and you have extended a hand to the future.

Finally, a word on your narrator. I am just a writer, not a scientist or a forester or an activist. I am knowledgeable about the natural world only to an average degree. If there is a story to how I came to tell this story, it is simply that some years ago I encountered the idea of trees being in the wrong places and thought that it was strange and fantastic and possibly important. It raised a series of questions about the present, past, and future arrangement of the world's living things. Once I started to think this way, it was hard to stop. I began to see trees in cities and parks and wonder where they'd come from, and who planted them, and why. I began to see dead trees and new seedlings out in the woods and wonder where the woods were headed. I began calling foresters and scientists and tree lovers of all makes to ask them unanswerable questions about the future. If you do this kind of thing long enough, it is inevitable that you will find your way to Connie Barlow.

The first time we spoke on the phone, Barlow was in Albuquerque, New Mexico, staying with friends, which is something she does almost constantly. She and her husband, Michael Dowd, are wandering preachers. They travel the country in their Dodge Sprinter van, speaking at churches and other community centers. Dowd, a former pastor, preaches nondenominational sermons about the religious meaning to be found in acceptance of Darwin and Wallace's theory of evolution. Barlow gives talks on climate change and on cultivating what she calls a "deep-time perspective." During our phone conversations, Barlow speaks faster than I can type notes. She has a Michigander accent and is prone to loud bursts of laughter. Her e-mails are their own genre. They are pages long, dappled with phrases in boldface, italics, or block caps, each designed to draw attention to this or that important point. Her style as a correspondent tends to the imperative. She begins paragraphs with "Notice:" or "Note:" or "See:" and usually includes long reading assignments. I often wake up to a half dozen new e-mails from Connie Barlow.

By the time we first spoke, Barlow had been arguing about the true nature of the Florida torreya's problem for more than a decade. A volunteer group she had founded, Torreya Guardians, had sent torreya seeds to Georgia, Alabama, South Carolina, North Carolina, Virginia, the District of Columbia, Kentucky, Tennessee, Ohio, Michigan, Illinois, Wisconsin, Massachusetts, Pennsylvania, New Hampshire, and Vermont. The trees were growing all over the eastern United States. Despite the seeming success of her project, Barlow was frustrated. She had hoped—just as her critics had feared—that more people would be following her lead, shipping rare trees north en masse. Instead, the scientists she'd expected to come around to her view were still mired in debate. They often cited her project as a bad example. Few average citizens seemed to show much interest in either the problems facing forests or in climate change more generally, both surely interesting topics. Meanwhile, everywhere she looked were species in need of human help. What about the Brewer spruce? What about the Rocky Mountain bristlecone pine? "AAAAHHHH!" she sigh-yelled into the phone. "I just go crazy."

But a person does what she can. At Sandia Peak, near Albuquerque, Barlow collected seeds from a grove of alligator juniper, another conifer that she thought needed help moving north. The trees' seeds are surrounded by a fleshy fruit. She guessed that they would be more likely to germinate after a trip through the bowels of a beast. The Sprinter van's next stop was almost two hundred miles north, in the town of Chama, near New Mexico's border with Colorado. There are other alligator junipers at that latitude, but none in the immediate area. One afternoon, she ate the seeds she'd collected. Most juniper berries taste bad, she said, but the alligator juniper berry is nearly delicious. "It's sweet," she said. "Doesn't have that resinous taste."

Was what Barlow did next the right thing to have done? Was it helpful to the trees? Was it helpful to future humans or to other organisms? Was it harmful? Was it too little, too late or too much, too soon? I can say, at least, that what she did was unusual. People move trees around all the time, but rarely with such intent, such determination, such dedication.

The next morning, she told me, she'd gone for a long hike. At first she was vague, but I pressed her for details, just to make sure I'd understood.

I had.

Yes, she said. "I am the woman who poops seeds."

When I told her I didn't think many people would be willing to do that, she laughed and laughed.

THEY SEEM TO BE IMMORTAL

More than any other living thing, trees define their surroundings. They break up the horizon, mark the trail, soften hard edges. You can pick them out one by one, a single thread, or as a texture, the whole cloth. Entire scenes can be conjured by the presence or absence of trees: the forest; the field; the house on the corner with the maple in the yard. A tree is a rooted thing, so a tree is also a place.

One fall morning a few years back, I joined a group of U.S. Geological Survey scientists on a hike in Sequoia National Park. Set high in the Sierra Nevada of eastern California, the park was so named because it is a place where giant sequoias grow. Even by arboreal standards, members of species *Sequoiadendron giganteum* carry unusual geographic weight. They are the world's biggest trees and among its oldest and tallest—all of the things people like about trees in general taken to an exaggerated end. Every year millions of people travel from around the world to see them. They give the strong impression of permanence. "I never saw a Big Tree that had died a natural death," John Muir wrote. "Barring accidents they seem to be immortal." But on the day of my visit, the trees seemed newly fragile.

The first storm of the season had swept through the night before and left a skiff of snow that fluttered off the trees, sparkling in the light. I followed the field crew through the park's Giant Forest grove, past pines and firs, oaks and incense cedars. A sequoia is an unsubtle thing, but there in the spare, open galleries of the park's forests, they had a way of appearing suddenly, as if you'd turned some invisible corner.

"You see this really bad one?" said one of the field crew members, pointing up the trail at a tree that had just come into view.

"Wow," said Nate Stephenson, looking up at it. "Wow."

The tree's trunk was red-orange, wide as a school bus tipped on its end, free of branches for a hundred feet. It looked carved from desert rock—eroded, furrowed, pocked with woodpecker holes, and squashed like a candle melted down over its stick. At its base it opened into a deep fire hollow—what foresters call a catface scar—that angled closed as though the tree was zipped together from the top. The snow at its feet was carpeted with golden-brown leaves, each a curl of overlapping pointed scales.

Stephenson, an ecologist, was the group's leader. He craned his neck to take in the tree's crown. Great clouds of foliage billowed from limbs that were themselves thick as trees. But where there should have been lush blue-green was more golden brown. Maybe the tree had been girdled during the last fire, leaving it vulnerable, Stephenson said. "It's probably going to die."

It was October 2014, three years into a statewide drought. That summer, the sequoias had started losing their needles. Foliage turned brown and fell away in clumps. Stephenson hadn't seen anything like it in his four decades at the park. There were no records of anybody else seeing it before, either. Reporters had been calling from around the world, asking if the sequoias were dying. Stephenson wasn't sure what to tell them. He was beginning to worry.

Sequoias are rare. Giants lost in a sea of lesser trees, they grow in just seventy-odd groves scattered along a narrow sliver of the Sierra Nevada's western slopes, between forty-five hundred and eight thousand feet in elevation, from Lake Tahoe in the north to Deer Creek two hundred miles south. Rarity means vulnerability to small accidents, to a new disease, to a sudden change in conditions. Despite the durability of individual trees, the species has always seemed vulnerable.

European-Americans have been trying to save them for 150 years. They first learned of sequoias in the winter of 1853, when a man named Gus Dowd stumbled into a grove of the trees while chasing a grizzly he'd shot and wounded. Dowd was a hunter hired to supply meat to a crew of men digging a water ditch for a gold mine. He was known among his fellows as something of a storyteller, according to one of those fellows, a man named Keen, who recorded the event in his diary. "One evening Old Dow [*sic*] told us he had found the largest tree in America," Keen wrote. "The Boys laughed at him but he swore it was so." Keen also noted that the Boys were sick of eating bear meat.

For years, there had been rumors of giant trees growing high in the Sierra Nevada, but few people believed them. (The local Native Americans, of course, had always known of the *wawona*, the big trees, and might have dispelled some of the mystery had anyone bothered to ask.) Dowd finally persuaded the crew's captain to accompany him back to the trees. Before long, the *Calaveras Chronicle* printed a story about Dowd's discovery, and wire services soon carried news of the species around the world. They called it the Sylvan Mastodon, the Enormous Vegetable Production, the Big Cylinder.

A few months later, a team of men traveled to the grove, carrying saws and axes. When they arrived, they found that their saws were too short and their axes too small, so they attacked the tree instead with pump augers, long hand-turned drills used to hollow out logs for water pipes. For twenty days they drilled hole after hole through the tree, then spent two days driving in wedges. On the twenty-third day, while the workers were away at lunch (or so the story goes), a gust of wind blew it over.

After the tree fell the men cut a two-foot-thick slab from its midsection, which they carted first to Stockton, then to San Francisco, where they displayed it at the corner of Bush and Montgomery. This made some people angry. "We hope that no one will conceive the idea of purchasing

Niagara Falls with the same purpose!" the editor of one magazine complained. Said another: "From this beginning, unless the Goths and Vandals are arrested in their work, the destruction of the incomparable forest will probably go on till the last vestige of it is destroyed."

At the time, this sort of protest was unusual. Tree hugging of any kind was rare. Millions of bison still roamed the United States. Chestnut trees still towered over Appalachia. Flocks of passenger pigeons still blotted out eastern skies. The first oil well had not yet been drilled. The West was still a frontier. It was possible to hunt for grizzlies in the mountains of California. Looking back, the world of the 1850s seems impossibly abundant. Why conserve when there was plenty? These sequoias, though, seemed worth saving. While more of the big trees were felled in the next decades for sideshows and timber, eventually people did save most of them.

But it can be hard to know the right way to save something. Early conservationists thought the trees were threatened not only by the ax but also by fire, so the sequoias' would-be saviors tried to protect them from that, too. It was a mistake. Fire had always burned through the groves, low and slow. The trees needed it to clear away competitors and make space for their seedlings. Without it the groves grew in thick, so that any fire now would be a big one, hot and fast, able to reach the sequoias' boughs and burn through their thick bark. People have worked since the 1960s to return fire to the groves, but many of them remain clotted, without new seedlings. Even as scientists like Stephenson worked to fix past errors, they had grown concerned in recent decades about the possible effects of a warmer climate. Then the drought came, and the sequoias started losing their foliage.

All morning, I followed the scientists through the woods. The field crew stopped and peered up at each sequoia, judging what proportion of its crown had died. They recorded the figure in a handheld computer: 0–10 percent; 10–25 percent; 25–50 percent; 50–75 percent; 75–100 percent. The catfaced sequoia was the mangiest, maybe, but there were many other sick trees. For Stephenson, it was a nagging worry. After nearly forty years at the park he knew many of the big sequoias as individuals. The trees can live for three thousand years. He'd always assumed that most of them would outlast his own brief time there.

He was worried, but curious, too. Ever since Dowd found the sequoias, people had wondered why they live where they do, hidden deep in the folds of the Sierra Nevada, in this place and only this place. Stephenson thought the dead leaves could be an important clue.

Understanding the sequoias' current placement means understanding how they got there. "Have they had a career," Asa Gray asked, "and can that career be ascertained or surmised, so that we may at least guess whence they came, and how, and when?"

It was 1872, at the meeting of the American Association for the Advancement of Science, held in Dubuque, Iowa. Gray was a prolific American botanist, the author of dozens of books and manuals, and hundreds of scientific articles. Soon after his speech in Dubuque, Gray planned to retire from both his position as association president and his professional duties at Harvard University, where he'd worked for more than three decades. He would take this last turn at the dais, he told his audience, as an opportunity to describe something he had lately been puzzling over.

He'd just returned from a trip out west, his first. As he'd traveled across the country via the newly joined Union Pacific railway, Gray had watched the forests give way to prairies, prairies to desert, and desert to pine-clad mountains. He ascended the Sierra Nevada's dry eastern flanks, then, finally, descended the western slopes, "which, refreshed by the Pacific, [bear] the noble forests of the Sierra Nevada and the Coast Range, and among them trees which are the wonder of the world."

Gray traveled to the Yosemite Valley in the company of a young writer named John Muir. They visited the Mariposa Grove of sequoias, which

Euro-Americans had found soon after Gus Dowd's Calaveras Grove. The trees were as magnificent as people said, Gray told his audience, but this was not what he wanted to talk about. What he was interested in was how they got there. "Were they created thus local and lonely, denizens of California only?" he asked. "Or, are they remnants, sole and scanty survivors of a race that has played a grander part in the past, but is now verging to extinction?"

For a long time, people had mostly assumed that if a plant existed in a place, then that plant was where it was supposed to be. "The great Author and Parent of all things, decreed, that the whole earth should be covered with plants," wrote Isaac Biberg, a student of the great Swedish categorizer Carl Linnaeus, in his 1749 tract *The Economy of Nature.* But God also wanted different places to have different climates, Biberg explained, so He gave different plants "such a nature, as might be chiefly adapted to the climate; so that some of them can bear an intense cold, others an equal degree of heat; some delight in dry ground, others in moist, &c."

To further explain the present ordering of the world's species, Linnaeus himself offered a slightly amended version of Genesis. The Garden of Eden, he wrote, had been on a mountainous, equatorial island, furnished with all existing animals and plants, each arranged on the slope according to its adaptations—the lichens at the peak, the palms near the shore, and, presumably, the Tree of Knowledge of Good and Evil somewhere in between. As the waters receded from this paradisiacal isle, Linnaeus wrote, the land itself expanded, and plants and animals went with it, each in its place. "Thus you have heard," he wrote, "with how exquisitely wise and attentive a care the great Artificer of Nature has provided that every seed shall find its proper soil, and be equally dispersed over the surface of the globe."

The only trouble with Linnaeus's paradisiacal isle was that the divine dispersal actually wasn't quite equal. Some places with similar climates were home to similar sets of species—some of the same species lived high in the Pyrenees and in sea-level Greenland, for instance—but some places with similar climates shared few species, if any. As Georges-Louis Leclerc pointed out in 1791, the only animals shared between the Old and New Worlds were those that lived in northern climes. There was no overlap at all between the two hemispheres' temperate and tropical fauna. In his 1807 *Essay on the Geography of Plants*, the Prussian naturalist Alexander von Humboldt wrote that in the five years he and the French botanist Aimé Bonpland spent studying the flora of South America, "we never encountered any European plant produced spontaneously by the South American soil."

Humboldt thought it was the plants that had shifted about, not terra firma. But he struggled to explain how species had crossed areas with conditions that didn't suit them—how those that delighted in dry ground had crossed moist areas, and vice versa. The most logical answer was that the conditions themselves had changed, and gradually enough for species to migrate in step. That could explain why a single species of plant might reach some of the places that would suit it, but not all—some barriers could not be crossed. Yet this idea of changing climate raised still trickier questions. "Would an increase in the intensity of the sun's rays have spread in certain periods tropical heat to the zones near the poles?" Humboldt wrote. "Do such variations . . . occur periodically?"

Soon naturalists argued that the world's climate had indeed changed, not once but many times. In the 1830s, Scottish geologist Charles Lyell wrote that conditions "are in a state of continual fluctuation, the igneous and aqueous agents remodeling, from time to time, the physical geography of the globe, and the migrations of species causing new relations to spring up successively between different organic beings." The world changed, and its inhabitants followed. The history of Earth, meanwhile, stretched from just six millennia to hundreds of millions of years and more.

But there was still the question of how, absent divine intervention, certain species came to be adapted to given conditions. In 1858, Charles Darwin and Alfred Russel Wallace provided an answer, having arrived independently at the same conclusion. In papers published side by side in the *Journal of the Proceedings of the Linnean Society*, they argued that life-forms evolve, with the slight variations between individual organisms of the same species making those individuals more or less likely to survive and reproduce. Useful variations accumulate in populations, while harmful variations are weeded out. As conditions change, formerly favorable traits might become burdens. Wrote Darwin: "The smallest grain in the balance, in the long run, must tell on which death shall fall, and which shall survive." By this evolutionary view, the present placement of species and even their present existence didn't reflect some underlying order. For individuals and species alike, life was chaotic, competitive, and often fleeting.

Finally all the pieces were in place. Where once the arrangement of the world's flora had reflected the whims of an unpredictable and unknowable deity, now it presented a series of puzzles that could, perhaps, be solved. As Asa Gray told his audience in Dubuque, Iowa, "Much that is enigmatical now may find explanation in some record of the past."

Fossil hunters had recently discovered the remains of trees resembling sequoias in Alaska, Greenland, Iceland, England, and parts of mainland Europe, he said. These were likely the direct ancestors of the living trees. The sequoias had been part of an ancient northern forest, a crown set on the world's brow. Then the world had cooled, forcing a retreat. From a great distance, over the millennia, you might see the forest sliding south as a single mass. But really each species made its own journey, tree by tree, seedlings springing up and maturing and casting their seeds a bit farther south. The species jostled among themselves, splitting around mountains and rivers, some able to cross arid or waterlogged places, others forced to go around. For every species, there were failed attempts where, perhaps, after following a river valley for miles and millennia, it came to a spot where its seedlings would not grow, and the descending cold overtook it. Over eons the world warmed and cooled and warmed again, and the trees followed as best they could. The sequoias were not isolated now because there was nowhere else on Earth that they could have survived, Gray said. Rather it was that, over the millennia, they'd hit dead end after dead end until it was finally just these few scattered groves, growing in the only suitable places they'd managed to reach.

Gray thought another dead end awaited. The sequoias could neither advance farther up the mountain, where they would be too exposed to winter storms, he said, nor retreat downslope, where the conditions would be too dry. He found few sequoia seedlings in the groves he visited. Sooner or later he thought the pines, firs, and incense cedars would squeeze the sequoias out through force of numbers. He told his audience that he thought the ancient groves couldn't survive much longer, and he made a prediction that now seems prescient: "A little further drying of the climate, which must once have been much moister than now, would precipitate [their] doom."

In the fall of 1875, three years after Asa Gray's speech, John Muir set off from the Yosemite Valley in the company of a mule named Brownie, heading south. Gray had explained the sequoia's location in a general way, but Muir was looking for specifics: "What conditions, favorable or otherwise were affecting it; what were its relations to climate, topography, soil, and the other trees growing with it, etc.; and whether, as was generally supposed, the species was nearing extinction." That is, why does the sequoia live where it does and not where it does not?

Although there was no name for it then, there is now an entire branch of science devoted to this type of question. It is called biogeography, and its practitioners, biogeographers. (As John H. Brown wrote in *Foundations of Biogeography*, it is a name and title none of the field's pioneers would have recognized. Humboldt, Darwin, Wallace, Gray, and those who followed mostly referred to themselves—more elegantly, perhaps—as naturalists. "Scientist" likewise wasn't a recognized vocation—the term was coined in 1834 by William Whewell and didn't come into popular use for some time after that.)

Modern biogeographers have many tools that their eighteenth- and nineteenth-century peers did not: genetic analysis, computer models, a wide range of sophisticated laboratory equipment. But the central question of the field—why things live where they do and not where they do not—remains a tricky one. To answer the question, biogeographers must look for patterns: What separates the places where the species occurs from the places where it does not?

Muir and Brownie traveled south, wending their way from grove to grove, across steep ravines, through untracked chaparral and endless pine forests. They dropped more than a mile in elevation to cross the Kings River—"Plants, climate, landscapes changing as if one had crossed an ocean to some far strange land," he wrote. Finally, months later, having split the last of his bread with Brownie, Muir reached the sequoias at Deer Creek, the southernmost grove. "A careful scrutiny of the woods beyond failed to discover a single sequoia, or any trace of its former existence," he wrote. "Now all that remained was to descend the range, and make a level way home along the plain."

He'd managed to pick out a few patterns. One was that the sequoias kept to themselves. Unlike individual pines or firs, which could sometimes be found far from others of their species, there seemed to be no similarly wayward sequoias. He noted, too, that toppled sequoias took a long time to rot away, leaving behind hollows in the forest floor when they finally did. He looked for these footprints beyond the edges of the groves and found none. "We therefore conclude that the area covered by Sequoia has not been diminished during the last eight or ten thousand years," he wrote, "and probably not at all in post-glacial times."

Another pattern was water. Asa Gray and others had pointed out that the sequoias were most often found next to running water or sodden meadows, which could be taken as a sign that they were most able to compete with their rivals only where there was extra moisture. But Muir, loath to admit any kind of weakness on the part of the "godlike" trees, disagreed. He thought the sequoias themselves caused the flowing water.

"The roots of this immense tree fill the ground, forming a sponge which hoards the bounty of the clouds and sends it forth in clear perennial streams," he wrote.

Of most concern to Muir was Gray's prediction of the sequoias' future. Muir agreed with the elder botanist's assessment of the northern groves, where the species seemed to be in decline, but he'd found a different situation in the southern groves. The sequoias there seemed to dominate the forests where they grew, and many young trees joined the old. The sequoia "is not yet passing away," Muir wrote. "Judging from its present condition and its ancient history, as far as I have been able to decipher it, our sequoia will live and flourish gloriously until A.D. 15,000 at least."

That is, if people let it. During his trip, Muir had seen hints of the destruction to come. He'd encountered sawmills in several of the groves, he wrote, "all of which saw more or less of the big trees into lumber."

In the late 1880s, the land offices in Stockton and Visalia were suddenly crowded with people filing claims under the Timber and Stone Act of 1878, which allowed the federal government to sell off large blocks of land for just dollars per acre. These claims were clustered around the Kings River sequoia groves—including one of the biggest groves of all, in Converse Basin. In March 1888, the *San Francisco Chronicle* revealed that the people who filed the claims were employed by Austin D. Moore and Hiram C. Smith, owners of thousands of acres of timberland in Washington State and a large lumber business in Stockton, California. Over the previous decades single trees had been felled for exhibitions, and logging companies had made occasional attempts on whole groves, but nothing like what Moore and Smith had in mind. The next month, they sent a team of surveyors into the Sierra. "Our plan," Moore said, "is to build a great flume."

In the spring of 1890, Moore and Smith completed two sawmills in the mountains. In the fall they finished their flume. This flume was a strange shade of the Roman aqueducts, a wooden trestle topped with a V-shaped chute. Pulses of water from a dam at the top sent sawn lumber sailing through this chute, from the mountains down to the foothills town of Sanger, where it met a railroad. In photographs, flume herders use crowbars to untie logjams. Solemn women lean against the flume as water flows by in a blur. The town of Sanger held a party to celebrate. "Come and ride in the big flume," read an advertisement. "Dancing, Barbecue, Speaking, Flume, Music, Refreshments." Board by board, the forest began to run down the mountain.

I hiked into Converse Basin on a recent summer night. The basin lies just northwest of Sequoia National Park, in Giant Sequoia National Monument. Much of the area had recently burned over, and in the evening I followed switchbacking fire roads through the blackened woods. I camped near the Chicago Stump, a big sequoia felled for another exhibit, and lay awake half the night after hearing a sound I was certain was a hungry panther.

I spent the next day walking the old roads. Some of them were freshly tracked, and I passed cattle stockades and young plantations of pine and fir. Farther in, many of the roads were abandoned, grown over with manzanita brush and fir seedlings. I saw few human artifacts—a rusty choker cable and the faded husk of a Tecate. It was only the stumps themselves that gave away the destruction that had happened there. They hid in the regrown forest, big and dark. Alongside many of the stumps lay the trunks themselves. The trees are brittle, another of sequoias' ironies—they're built to stand for millennia but not to survive the fall to Earth. The loggers sometimes made beds of lesser trees to catch the giants. Even so, many of them shattered. The biggest trees were turned into little things—slats and shakes and shingles. The old heartwood is slow to rot. A century on, the broken boles were hardly diminished.

Kicking aside pinecones, pursued by little sweat flies, I followed the dusty logging roads to a bare ridgeline. I could see the snowcapped Sierra

Nevada stretching north and the smudge of the Central Valley to the west. Then I descended, arriving at what is now called Stump Meadow as the Sun set. The stumps and wasted logs lay together like the ruins of some ancient temple. In the end, the sequoias were too big, too difficult, too far from coastal markets. In 1895, after investing more than a million dollars in the venture, Moore and Smith filed for bankruptcy. Two years later, new owners extended the railroad into the middle of Converse Basin, where they built a new mill. The operation changed hands a third time. It wasn't until the 1920s that logging in Converse Basin finally ceased. I sat on a stump and had my dinner and listened to the frogs. It was a somber place.

Despite the loss of one of the biggest groves, the people on the other side were winning the fight to save the species. Grove by grove, they fundraised, lobbied, and argued, building parks and preserves to hold the trees. Certain people always become shorthand for the things they are a part of, and Muir suits to sum up the accomplishments of his time as well as any. When he died, on Christmas Eve 1914, the *New York Times* hailed him as the "Guardian of the Yosemite" and the "Naturalist of the Sierras," and noted his many important friends and acquaintances, among them former president Theodore Roosevelt, writer Ralph Waldo Emerson, and inventor Thomas Edison. Muir had helped found the Sierra Club, aided in the creation of several national parks, and written dozens of popular articles and best-selling books.

But Muir often told his own story as one of loss. As a boy growing up in Scotland, he'd read schoolbooks about the strange and vast American wilderness—the bald eagles, the trees that dripped sugary sap, the endless clouds of passenger pigeons. Soon after the family arrived in Wisconsin, when Muir was twelve, he saw the pigeons. "Of all God's feathered people that sailed the Wisconsin sky, no other bird seemed to us so wonderful," he wrote. In the fall, the pigeons would arrive in flocks so big that it took them all day to pass over, "like a mighty river in the sky." Muir often joined in on pigeon hunts. "Every shotgun was aimed at them," he wrote, "and everybody feasted on pigeon pies." Recalling the experience decades later, Muir knew that the birds, once symbols of the continent's boundlessness, had grown rare. He was furious. "Think of the millions of squabs that preaching, praying men and women kill and eat, with all sorts of other animals great and small, young and old," he wrote, "while eloquently discoursing on the coming of the blessed peaceful, bloodless millennium!"

Martha, the last of the passenger pigeons, died the same year as Muir. By then, the buffalo herds and the Californian grizzlies were nearly gone, too. The western frontier was long closed, and the world had begun to run on oil. Chestnut trees still stood, but not for long. The previous winter, President Woodrow Wilson had signed a bill allowing the damming of Hetch Hetchy Valley, just north of Yosemite Valley, to provide water to San Francisco. Muir had fought the dam. Some said he died of a broken heart.

Later, a draft titled "Save the Redwoods" was found among his papers. "God has cared for these trees, saved them from drought, disease, avalanches, and a thousand storms," Muir wrote. "But he cannot save them from sawmills and fools; this is left to the American people." Gradually the people did save the sequoias. By the 1950s all but a few groves were publicly owned, walled off from the ax and protected from fire.

But this kind of preservation is that of the mummy, of formaldehyde, of a shuttered room—the idea that in mere separation lies salvation. The mummy turns brittle, the specimen shrivels and discolors, the room grows stagnant. Brush sprang up, and other trees, and over the years the wide galleries between the giants grew crowded.

During his 1875 trek south, John Muir spotted a fire just outside a grove of giant sequoias. The fire swept upslope through chaparral, "a broad cataract of flames," he wrote, "lurid flapping surges and the smoke and the terrible rushing and roaring." But then the fire hit the sequoia grove and "became calm, like a torrent entering a lake." It crept along, low to the ground, "slowly nibbling the cake of compressed needles." Muir described the sequoias' ability to weather the fires behind their thick bark, and the regularity of these fires, every five or ten years, whenever enough leaves and debris had built up to carry a flame. He observed that the fires cleared duff and shrubs, providing the bare soil that the sequoias' seeds needed. But Muir thought the seeds would do as well in clearings left by wind-felled trees. He feared fire's effects, calling it the "great destroyer" of sequoias, damaging to the old trees and fatal to the young. Up until his death, he begged in his books and editorials that the sequoias be protected from the flames.

Muir's view of fire was a common one at the time, based on old ideas of forestry from Europe and reinforced by a series of recent disasters in the United States. Industrial logging in the East and the Midwest left behind forests filled with debris, primed to burn. In one of the worst catastrophes, in 1871, a forest fire killed more than six hundred people in the town of Peshtigo, Wisconsin. In the public discourse of the day, fire was seen as destructive, not life-giving. Although some early grove managers allowed fire to burn through the sequoias, for most of the first half of the twentieth century, official policy was to keep fire out.

But after the fires stopped, no more new sequoias sprouted. Shade-tolerant species grew in, and the ground below became cluttered with branches and duff. Instead of the low, slow-burning fire Muir had witnessed, a fire now would be a conflagration, capable of killing even the giants. Muir's fear became self-fulfilling: Protecting the groves from fire made them vulnerable to fire.

By the 1950s, the damage was apparent. A number of ecologists began to push for the reintroduction of fire. In 1963, the National Park Service published what is now known as the Leopold Report, after its lead author, A. Starker Leopold. The report outlined what was to be the Park

Service's primary goal for the next five decades: restoring the parks to their condition before the arrival of European-Americans. "A national park," the report's authors wrote, "should represent a vignette of primitive America." In some parks, this would be impossible, they wrote, as in the eastern United States, where the hardwood forest could regrow, "but the chestnut will be missing and so will the roar of pigeon wings." Other parks, though, were just overgrown.

In the fall of 2017, I stood at a T in the road in Kings Canyon National Park and watched as a crew of firefighters lit a hillside on fire. Flames licked up the side of snags and rushed through whitethorn and manzanita brush. Smoke rose across a blue sky. Ash fluttered down and landed on my notepad. A firefighter in a yellow coat, green pants, and mirrored glasses stood with his back to the fire, watching over a line of idling cars. Behind him, other firefighters meandered through the woods. They waved drip torches, thermos-shaped jugs with curly pig's-tail spouts that dribbled flaming liquid, leaving little burning slug trails that caught and spread through the scrub.

Grove managers started lighting fires here in Kings Canyon National Park and its sister park, Sequoia, in the late 1960s. Nate Stephenson, the USGS ecologist I joined on the field survey in Sequoia National Park, first worked at the park in 1979, doing odd jobs, including a stint as an emergency firefighter. He was hired a decade later as a fire scientist. Back then, prescribed fire was seen as a panacea, the only real help the sequoias needed, he said. By the mid-1990s the prescribed fire program was two decades old and was considered the best in the western United States. Restored this way, the vignette might last for centuries.

But some of Stephenson's colleagues ("smarter than me," he said) wanted to know how closely the program was actually matching the frequency and extent of fire in the pre-Euro-American era. Park Service fire ecologist Tony Caprio tallied fire scars on the sequoias' annual growth rings to estimate past fire frequency. Caprio and his colleagues compared fire records from different parts of the forest to estimate what portion had burned each year. They found that the grove managers were burning less than a tenth of what had burned annually before the fires stopped.

At the prescribed fire in Kings Canyon, the burn bosses I talked with told me they have to thread a logistical needle, simultaneously gaining approval from air quality managers, catching the weather conditions that will promote a fire that carries without growing too powerful, and securing both the money and the people to carry out the work. The fire crews I found lighting fires had spent the summer fighting wildfires across the West. Runaway fires are a worry, too. Burning any more would be difficult, Stephenson thought. Matching historical rates would be nearly impossible.

Outside the park, the situation was even worse. In 1996, Stephenson published a chapter in a report to Congress that summed up where things stood, a sequoical State of the State. People had reintroduced fire to less than a fifth of the sequoia groves, he wrote. Another fifth of the groves had been logged (among them Converse Basin). The remaining three-fifths had been protected from both logging and fire, leaving them vulnerable to catastrophic blazes and without offspring. He realized that the sequoias still hadn't been saved.

Two years after our hike through the Giant Forest with the field crews, I met Nate Stephenson at the Sequoia National Park U.S. Geological Survey field station. He is lanky, with Abe Lincoln's strong chin and a scruffy, graying beard. He offered coffee. He drank his from a tiny mug his wife had given him so he'd get up from his desk more often. Then we drove the winding road up the hill, past tan oaks and black oaks, buckeyes and bay laurels, incense cedars, pines, firs, and finally the sequoias. We parked in a lot and headed into the woods.

Walking with Stephenson through the park's Giant Forest grove felt like getting a tour of someone's house. He seemed to know every tree, every bend in the path. We passed a tattered little sequoia that he said was much regarded by the local chickaree squirrels, who lined their nests with its fibrous bark. They seemed to prefer that particular tree over all others. We paused so he could teach me how to tell the difference between white fir and red fir, a subtle thing. "I've come to appreciate red firs," he said. "I used to think, 'Ehh.' But I've matured."

Farther on he stopped in front of his favorite sequoia. It was hollowed out by fire on two sides, its charred center open to the world for a hundred feet, but above was boisterous green. "If I can wax philosophical for a moment," he said, "I've been coming by this particular sequoia for thirty-eight years . . . and it hasn't changed. I think that's a reason a lot of people come to see the sequoias. It seems like something permanent in this crazy world." He put his hand on the tree and leaned back to look at it. "It's had the shit beat out of it," he said. "But it just keeps chugging along."

For a long time, he struggled to find things in the park to feel hopeful about. When he realized how little the prescribed fire program was actually doing, it was a "crashing epiphany," he told me. "You're in denial, and you don't even know you're in denial, and suddenly you can't maintain the denial anymore." The possible effect of climate change on the trees, previously a distant problem, suddenly overwhelmed him. "I found it very depressing," he said.

In 2013, ecologist Richard Hobbs wrote that "people with an interest in species, ecosystems, and the environment in general are constantly assailed with accounts of past or impending loss." People mourn the diminishment of the natural world the same way they mourn the death of a family member, Hobbs wrote. As psychiatrist Elisabeth Kübler-Ross argued in her 1969 book, *On Death and Dying*, a person experiences grief in distinct stages: first denial, then anger, bargaining, depression, and, finally, acceptance.

We reached a small stream, and Stephenson suggested we stop for lunch. ("I have a shrewlike metabolism," he said.) He sat cross-legged and pulled out a peanut butter sandwich and an apple. It took him a long time to emerge from his gloom, he told me, nothing like the steep drop into it. Denial, anger, bargaining, depression. Slowly he admitted to himself that Muir was wrong, that the authors of the Leopold Report were wrong. The sequoias couldn't stay here, unmoving, to A.D. 15,000. They might not even make it to A.D. 2500. Now he is nearing retirement. His generation hadn't saved the sequoias, but there are still things he wants to understand better, so that maybe in the future, another generation will save them. "I'm through grieving the loss of an ideal," he said. "Now it's 'What are we going to do about it?' " The central question, still, is why sequoias live here and no place else.

"Niche" was originally an architectural term, describing a small, narrow hollow in a wall where a statue or piece of art could be displayed. By the eighteenth century the word had come into use in the modern, popular sense: "a place or position adapted to the character or capabilities, or suited to the merits of a person or thing," as defined by the *Oxford English Dictionary*. In 1917, biologist Joseph Grinnell used the word in a new way, as an ecological concept, to describe the habitat of the California thrasher, *Toxostoma redivivum*, a small taupe bird with a sickle-shaped bill. "Its range is determined by a narrow phase of conditions obtaining in the Chaparral association, within the California fauna, and within the Upper Sonoran life-zone," Grinnell wrote. "The nests of the Thrasher are located in dense masses of foliage, from two to six feet above the ground, in bushes which are usually a part of its typical chaparral habitat." Ten years later, ecologist Charles Elton offered another version of the ecological niche, defining it as the role a species plays in an ecosystem. "For instance," he wrote, "there is the niche which is filled by birds of prey which eat small animals such as shrews and mice. In an oak wood this niche is filled by tawny owls, while in the open grassland it is occupied by kestrels." As people later summed up the two definitions, Grinnell's niche describes a species' biological street address, while Elton's niche describes a species' biological occupation.

There is a more interesting and useful definition. In this version, proposed by biologist Evelyn Hutchinson in 1957, the niche is not something an organism fills or does but is instead internal to the organism. This niche contains all the possible combinations of conditions in which a population of a given species might survive, grow, and reproduce in perpetuity—all the iterations of *climate + habitat + food + et cetera* that the species might bear. Hutchinson called his version the "fundamental niche," and he described it as a multidimensional "hypervolume," a shape of many sides. Theoretically, in the very center of that space exists the species' utopia, the part of the paradisiacal isle that perfectly combines all the conditions that would allow the species to live its best life. Most of the hypervolume is not perfect for the species, though—merely survivable. This survivable space fades toward the hypervolume's edges until finally survivable combinations of conditions give way to unsurvivable combinations. To perceive the entire fundamental niche of a species would be to know its every side. But while it is possible to find certain edges, other facets remain hidden. Hutchinson's fundamental niche is less like a species' street address or job and more like its soul.

It's possible that giant sequoias are already living in their paradise. On the whole, though, the evidence suggests that the trees are sitting close to one of the edges. But what is that edge? What is it that separates the places where sequoias grow from where they don't? The likely answer, Stephenson said, is the one that's been most obvious all along: water.

As Muir and others noticed, what seems to divide places with sequoias from adjacent places without sequoias is the presence of groundwater. But Muir had it backward, Stephenson said. The big trees don't cause flowing water. They grow where they do because of it. California has a Mediterranean climate, receiving much rain and snow during the winter and almost none the rest of the year. Groundwater means that the sequoias are spared much of this dry stretch. But these wet patches are fragile.

They're fed by melting snow from farther upslope. Already, spring comes earlier to the southern Sierra Nevada, and the snow melts faster than in centuries past. The dry season is getting longer.

By the late 2000s, the parks' scientists had begun focusing much of their attention on the possible effects of climate change, Stephenson said. They knew that ten thousand years ago, when the Sierra Nevada were warmer and drier than today, the sequoias were even rarer. They huddled at the very edges of streams, teetering on the brink of extinction. Then, about forty-five hundred years ago, the Sierra Nevada grew a little cooler and wetter, and the trees spread out into their current distribution, reaching their modern abundance just two thousand years ago. But it wasn't clear how the warm, dry Sierra Nevada of the past compared with the warmer, drier mountain range of the future, or what would happen during the transition to that climate.

"We were grappling with these questions of what the future might look like," Stephenson said, "and then Mother Nature comes along and says, 'Ah, I'll give you a preview.'" In the winter of 2012 little snow fell in the Sierra Nevada, and even less the winter after. The next year, 2014, was California's driest year in at least a century. That summer the sequoias began losing their leaves. To a sequoia researcher, it was fascinating, Stephenson told me, but it also made him nervous. Over the next two years, though, the drought slackened, and the trees shed fewer leaves. In 2017 heavy snows returned. The big catfaced sequoia died, as Stephenson had predicted it would, but few others did. By the summer of 2018, the sequoias flushed with new foliage.

Stephenson and his colleagues now think that the dead foliage might have helped the sequoias survive. Trees are essentially bundles of straws. Below the outer bark, which is dead tissue like hair and fingernails, lies the phloem, a layer of straws that carry sugars down from the leaves. Below the phloem is a thin layer of cells called the cambium, where the tree's cells divide and produce new tissue. Below the cambium is the xylem, which carries water and minerals up from the roots. In photosynthesis, leaves use the energy from sunlight to convert water and carbon dioxide into sugars, with water and oxygen as by-products. This flow of water from the leaves into the air creates a negative pressure in the straws in the xylem, pulling water up from the roots the same way you suck up a milkshake. When the leaves transpire water out faster than the roots can supply it, the suction force in the straws increases. If the leaves suck too hard, the water column in some straws will eventually break apart, or "cavitate," cutting off the flow of water and shutting down photosynthesis and growth. If enough of the straws cavitate, the plant will die.

The dead leaves reduced the amount of water the sequoias needed, perhaps preventing worse damage. But growing new leaves is costly, too, and the trees missed out on part of a growing season. So far as we can assign intent to a sequoia, shedding bunches of foliage looks like a desperate move. The data Stephenson and his colleagues collected seem to confirm this conclusion: The trees in drier and steeper sites lost a greater share of their foliage, as did the trees at the edges of the groves. It is as if the boundaries of the sequoias' little refuge was pushing in on them.

For now, the danger seems to have receded. The sequoias' minders are biding their time, watching and studying, trying to better understand the trees' history and their possible future. They'll keep them here as long as they can. But sooner or later, climate change will threaten the sequoias, Stephenson said. "I think we can say with fair certainty that if it gets warmer and warmer and warmer and warmer, as the projections have it, at some point we're going to see the sequoias suffering." Between 1985 and 2015 the average temperature in the sequoia groves of the southern Sierra Nevada increased by about four degrees Fahrenheit. The hotter it is, the more water plants transpire. Even if the same amount of snow and rain falls in the future, even if there are no more long droughts, a warmer climate will mean more water out, and therefore more stress on the trees.

The real problem might be the seedlings. Earlier in the day, Stephenson had picked up a sequoia cone. It was the size of a golf ball, a green-brown Fibonacci spiral of diamond-shaped, puckered-lips segments. He banged it against his palm, knocking out little black grains—tannin, he said—and a single seed. The seed was yellow-orange, the size and shape of a rolled oat, a flat oval with a line down the middle. It was a ridiculous little seed. "Not one seed in a million shall germinate at all," Muir wrote, "and of those that do perhaps not one in ten thousand is suffered to live through the many vicissitudes of storm, drought, fire, and snow-crushing that beset their youth." The sequoias' current locations reflect where the seedlings could survive. Though they are small enough to be carried away by wind, the seeds find no purchase outside of the groves.

Maybe this is how it goes: The mountains get warmer and the little wet places where the sequoias live get drier, and fewer and fewer seedlings survive. There is a long drought, and then a few decades later, another. Slowly the groves shrink, as the trees at the margins lose their leaves and don't recover. For a time, perhaps, the grove managers are able to water the groves, but eventually it becomes too hot. The trees have hit another dead end.

Conservation of the type championed by Muir—sequoias as wax museum—is out, as is his understanding of the "Nature" he encountered, separate from and superior to everything of and by humankind. The man himself gets a critical second look. As Eric Michael Johnson wrote in *Scientific American*, Muir was disdainful of the Ahwahneechee people he encountered in the Yosemite Valley, whose fires had helped sculpt the valley's parklike forests. His eloquent writing about an unpeopled wild at times helped justify the removal of Native Americans from their lands. "John Muir's idea of conservation is still affecting indigenous people throughout the world," said Irene Vasquez, a member of the Southern Sierra Band of Miwuk (as yet federally unrecognized) and a descendant of Yosemite's Ahwahneechee inhabitants. Archaeological evidence shows that people have lived in the Sierra Nevada for many thousands of years. People had likely watched the tree that Gus Dowd "discovered" in 1853 grow from a seedling. Set against this long history, Euro-American efforts to protect the sequoias—*wawona* in the Miwuk language—look even more disastrous. "Indigenous people knew their lands," Vasquez told me. "They knew how to take care of them."

The Leopold Report is out, too. Like Muir, its authors imagined a pre-Columbian America that didn't exist. In 2012, an advisory board delivered a report to the National Park Service's director, Jonathan Jarvis, titled "Revisiting Leopold." They advised that the service's "management strategies must be expanded to encompass a geographic scope beyond park boundaries to larger landscapes and to consider longer time horizons." The national park was recast as a "moving record of a dynamic and continuously changing system." The old ideas of conservation rested on underlying stability. They assumed that not everything would change at once.

After lunch, Stephenson and I walked back up into the forest. Along with a few sequoias there were big pines, firs, and incense cedars, and many younger trees, as well as standing dead trees and others lying crisscross on the ground in various states of decomposition. The sequoias are icons, charismatic megaflora, and during the drought they were what attracted the media's attention (including mine). But the sequoias are a relative few. "All my observations go to show that in case of prolonged drought the sugar pines and firs would die before Sequoia," Muir wrote. It seems that he was right.

Stephenson paused near a small, dead fir and pointed out an aluminum tag hanging on a nail. The tag was punched with the number 189. Stephenson's colleague Adrian Das later sent me the full report: Found dead in 2016; weakened by root rot, killed by beetles. The geological survey keeps files like this for every one of the thirty thousand–odd trees growing in thirty plots scattered around the park. Every summer, field crews measure every tree, noting its condition and diameter. If a tree has died, they give it an autopsy.

Stephenson helped tag some of the original plots back in the early

1980s. The program is now the longest-running annual monitoring program of its type, he told me. Checking in on every tree every year has allowed the park's scientists to catch things they'd miss with a less regular schedule. The length of the program has provided an unusually full picture of the forest. Some discoveries seem trivial: A dead tree has a vertical half-life of approximately eight years—that is, a 50 percent chance it will topple over within eight years. Others are troubling: During the three-year drought, 70 percent of large sugar pines in some plots died—a blink of the eye in the life of a species that can live five hundred years. More troubling still: Even before the drought, the number of trees dying in the plots each year was rising.

Stephenson stopped in front of another dead fir. "Oh man, beautiful *Scolytus* galleries!" he exclaimed. When the field crew gave the tree an autopsy, they removed an eight-inch patch of bark with a drawknife, opening a window that Stephenson now peered through. Between the bark and the wood was crumbled brown material like clotted blood where the tree's life used to flow. A horizontal channel the width of a spaghetti noodle ran across this dried phloem. A half dozen smaller channels hung from the first in an upside-down candelabra. The horizontal line was where the mother fir engraver beetle, *Scolytus ventralis*, tunneled through and laid her eggs. The tracks below recorded her larvae's subsequent feast, the lines growing bigger as the grubs wriggled deeper into the phloem.

Each dead tree died a little differently, but together they told a single story. On the way up from the USGS field station at Three Rivers, Stephenson and I passed scattered dead oaks, hillsides of pines gone rust red. The drought had killed some 150 million trees across California. The southern Sierra Nevada were hit the worst.

A tree defines its place, so when the tree dies it can seem like the place is ruined, too. As Stephenson pointed out all the dead trees in the long-term plot, it felt a little like the end of the world. But I tried to remind myself that what is happening now is what has always happened: The climate changes, and forests follow.

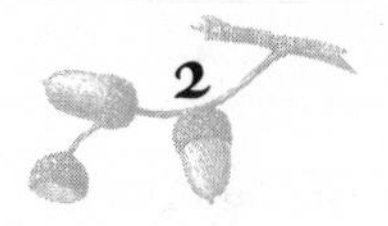

THE HOLOCENE

One summer morning in 2018, I followed Ed Berg down a concrete boat ramp in Homer, Alaska. He carried a walking stick, a piece of old driftwood with a length of orange flagging tied around its middle. Berg was weatherproof: rain pants, rain jacket over a Shetland sweater and an orange utility vest, and the type of brown rubber boots that are ubiquitous in Southcentral Alaska. He is soft-spoken, with white hair, blue eyes, and a beard—a kind of seafaring elf.

Berg eased himself over the riprap boulders at the ramp's edge. The beach was black sand and cobbles, strewn with tangles of sea gump and logs as rounded as baby carrots. The water was out, way out—there the tide is a wild pendulum, swinging fifteen feet or more. Bare mud shone for a quarter mile. Across Kachemak Bay, mountains rose straight from the water's edge. Dark clouds threatened more rain.

It was a fast, jagged landscape, but a slower, gentler one lay hidden just barely out of sight. Berg, a retired ecologist and geology professor, was attuned to its signs. Here, he pointed out pale boulders protruding from the sand—concretions, he said, spots where flowing water eddied and dropped suspended grit, now welded together by the ages. And there, black ledges that stuck out from the sand at shallow angles—coal seams, he told me, the remains of old peatland, carbonized. The coal is low-grade, sub-bituminous, but there's plenty of it. The people in Homer sometimes use it to heat their homes. It burns hotter than wood and has been known to put holes in cast-iron stoves.

These concretions and coal seams are the tracks of an ancient river, Berg said. It was there before the mountains rose, before the sea intruded. The river swept back and forth through peatland, and in the driest parts of the floodplain, trees grew. Now, down well past the tide line, we came upon one of them—or, rather, the stump of one. It was fossilized, crusted with barnacles and mussels and sea plants. It looked a lot like the surrounding rocks. The years had been chewing on it. It was one of the ancient trees Asa Gray described in his 1872 speech—not a *Sequoiadendron giganteum,* like the trees in Sequoia National Park, but a member of a related genus that scientists now call *Metasequoia*—"like sequoia." Berg poked it with his stick. Just down the beach were several others. He first came across them in the 1980s, he said. I asked him whether he discovered them on his own, or if someone told him where to look. He chuckled. "Did I discover them? Boy, I don't know. I don't remember anyone showing them to me. Okay. I guess I did discover them. But I'm sure other people noticed them." I think he was being modest.

At the time Berg's tree lived—mid-Miocene, he said, about 10 million years ago—the sequoias and their relatives were among the commonest trees in the world's northern latitudes. Often their remains can be found alongside those of maples, poplars, oaks, beeches, magnolias, hickories, liquidambars, birches, pines, and—as Gray put it, summing up a similar list—"whatever else characterizes the temperate-zone forests of our era."

It's a mismatch: the fossils of trees that resemble modern trees, but far north of where the modern trees live. The fossils are what scientists call a "proxy," a small hint of a scene that is otherwise invisible. Ed Berg's concretions and coal seams are proxies for a long ago floodplain. The growth rings that trees lay down each year are another kind of proxy. Relatively thicker rings mean more favorable growing conditions; relatively thinner rings, less favorable. The fossil metasequoias, meanwhile, are proxies for an entire vanished climate. The trees of the Miocene forest are similar enough to the trees of modern temperate forests that scientists can use their knowledge of where the modern species live to learn something about the place where their fossilized counterparts lived, too. "We infer the climate from the trees," Gray wrote, "and the trees give sure indications of the climate." When the sequoias lived in the Arctic, he said, Alaska had been as warm as New Jersey.

Then the world grew colder. After the Miocene epoch came the Pliocene, and then the Pleistocene. In 1837 the Swiss-American naturalist Louis Agassiz made a startling claim: During this last stretch of time, which immediately preceded the rise of human civilization, vast sheets of ice had covered large parts of the world. (Others before Agassiz had suggested the existence of a recent ice age, but Agassiz deserves credit for popularizing the idea.) Hints of these ice sheets were readily visible around the world, he wrote, if one knew what to look for. He pointed to moraines (the long furrows of crushed rock shoved up by ice), to so-called erratic blocks (enormous boulders that seemed to have arrived out of thin air), and to strangely polished and lined mountainsides.

In an 1859 article on the flora of Japan, Asa Gray imagined the scene at the beginning of this ice age. Plants "must have been pushed on to lower latitudes as the cold advanced," he wrote, "and between them and the ice there was doubtless a band of subarctic and arctic vegetation." Then, as the ice retreated, "the surviving temperate flora must have returned northward *paripassu*"—in step, a dance between plants and ice.

Every species has the capacity to move from one place to another. The individuals of many species move around during their lifetime, migrating with the seasons, as birds do. Other species, like monarch butterflies, might take several generations to make a round trip. For immobile, or "sessile," creatures like trees, migration works differently. It is both constant and communal, accomplished not by individuals but by populations or even whole species.

Trees cast off seeds in every direction, sending them whirling down on paper wings, stuffed in a chipmunk's cheek, or floating away in a buoyant husk. The first and only journey of an individual tree is made at random. The species, in turn, is steered only by the fact of where each seed survives and where it doesn't, by the successes and failures of a million aimless journeys. When conditions remain steady, seeds succeed and fail in roughly the same places. The species is held in place. When conditions change, seeds succeed and fail in new places. The species moves.

In his 1859 article, Asa Gray argued that such movement could explain many of the mysteries of floral geography—why, for instance, similar species sometimes appear in widely separated parts of the world. It was clear, he wrote, that species do not remain in their place of origin, but roam about. Their present arrangement reflects what has happened to them along the way. As scientists grew in their ability to decipher the proxies of past ages, they found ample evidence that Gray was right. The fossil record shows that species have ranged widely, often arranging themselves together in ways unlike the present. In periods of change, rare species have grown suddenly common, and common species suddenly rare. Other times, when the climate has changed, species have not been able to keep up.

Fossils that did not match the places where they were found showed that Earth had changed, but naturalists of the eighteenth and nineteenth centuries disagreed about how it had changed, and why. In his 1813 *Essay on the Theory of the Earth,* French naturalist Georges Cuvier suggested that the planet had suffered a series of worldwide floods. "Numberless living beings have been the victims of these catastrophes," he wrote. "Some have been destroyed by sudden inundations, others have been laid dry in consequence of the bottom of the seas being instantaneously elevated." Cuvier didn't think that the geological forces at work in the world today could account for the changes visible in the fossil record. "The march of nature is changed," he wrote. Whatever had upended the planet before, it couldn't happen again.

Decades earlier, in his own *Theory of the Earth*, Scottish geologist James Hutton had argued to the contrary. He thought that all the changes suggested by fossils and landforms could be explained by processes still visible: erosion and deposition, earthquakes and volcanism. Change, Hutton wrote, came not catastrophically, but gradually and steadily. The key ingredient was time—lots and lots of time. He thought the world's layered rocks held evidence that Earth was unfathomably old. "We find no vestige of a beginning," he wrote, "no prospect of an end."

In the 1830s a second Scottish geologist, Charles Lyell, expanded upon Hutton's "Uniformitarianism" and argued against Cuvier's "Catastrophism." (Both terms were coined by William Whewell, the British scientist who also came up with the word "scientist.") Lyell thought that the world's climate depended on the arrangement of its continents. The current position of the continents—with Asia, Europe, and North America huddled around the Arctic Ocean, and another landmass squarely Antarctic—prevents warm, tropical water from mixing with cold, polar water, thereby making the world cooler than it would be if the continents were set around a more equatorial center.

As climatologist Mark Maslin wrote in *The Complete Ice Age*, Lyell was correct. By 100 million years ago, the arrangement of Earth's continents made its climate sensitive to slight changes in the concentration of atmospheric gases and in the planet's orbit around the Sun. Antarctica grew an ice sheet around 35 million years ago but shed it 10 million years later. After another 15 million years, the Antarctic ice sheet formed again, and not long after, Greenland grew an ice sheet, too. At 2.74 million years ago, ice sheets began to spread across northern Europe and Asia, and at 2.7 million years ago, an ice sheet began to form in North America. White ice reflected more of the Sun's energy, and the temperature fell further. The Pleistocene began.

During this epoch, the ice advanced and retreated dozens of times in stretches called glacials and interglacials. These are caused in large part by three overlapping cycles—of 96,000 years, 41,000 years, and 27,000 years—that affect when sunlight reaches Earth. The first cycle is a change in the shape of Earth's orbit around the Sun, from more circular to more elliptical. The second cycle is a change in the tilt of the planet's axis of rotation. The third cycle is a change in the wobble of Earth's axis, like a spinning top as it loses momentum. These Milankovitch cycles (named after Milutin Milankovitch, the Serbian physicist who first proposed them) do not change the total amount of solar radiation that reaches Earth, wrote E. C. Pielou in *After the Ice Age.* Instead, they sharpen or dull the difference between summer and winter. When the difference is slight—when winters are mild and summers are cool—the heat of summer is not enough to melt the winter's snow. Snow accumulates, and glaciers form.

In North America, the most recent glaciation is called the Wisconsin. When it waxed, 21,000 years ago, the ice sheet that covered North America was as big as the one that now caps Antarctica. It covered nearly all of Canada and much of the northern United States. Parts of it were two miles deep. The ice was so heavy that it depressed Earth's crust, the way you might pinch a tennis ball between your thumb and forefinger. In places, the crust is still rebounding.

Then, around 18,000 years ago, the heat of summer grew strong enough to melt winter's snow. The ice began its retreat. By the beginning of the Holocene epoch, 11,700 years ago, the ice had receded as far as the present Great Lakes. Maine and New Brunswick were free of ice, and a causeway of bare ground extended to Alaska. By 8,400 years ago all of western Canada and Alaska was free. By 5,000 years ago only a few scattered remnants of the ice sheet remained, on Baffin Island and huddled here and there in the mountains. Across a newly open continent, trees sprouted.

As with city planning departments, the tendency of time is to tear down what came before. Darwin compared the fossil record to a book: "Of this volume, only here and there a short chapter has been preserved; and of each page, only here and there a few lines." Ancient trees that remain rooted in the place they grew, like the metasequoias of Homer, are rare. But subtler hints of ancient forests can be found all over, if you know how to look for them.

On a summer day in 2017, I joined a group of U.S. Geological Survey scientists on a trip to a place they call the Donut, a few hours south of Anchorage, Alaska, on the Kenai Peninsula near Funny River. The Donut is a circle of black spruce growing in peatland. Peatland is the broadest term for this type of mossy wetland. Miriam Jones, the expedition's leader, told me that the wetland that surrounds the Donut is specifically a fen, which is a mossy wetland that is made wet partly by groundwater. The inside of the Donut is a bog, which is a mossy wetland that receives its water primarily from the sky. Bogs and fens tend to be acidic and oxygen-starved. Instead of rotting away, organic material accumulates.

Pursued by a cloud of mosquitoes, we slogged single-file down from the road across several hundred feet of fen, which threatened to fill our hip waders. The carpet of sphagnum moss was overlaid with cranberries and blueberries, crowberries and rushes. Trees grew in the parts of the fen that were not flooded but merely sodden. The Donut was one of these. It was about a hundred feet in diameter, a circle of forested ground underlaid by permafrost—ground that is permanently frozen. Frozen ground is bulkier than unfrozen ground, which leavened the Donut and made it a good place to drop our bag and drain our hip waders. The center of the Donut, meanwhile, was jelly-filled. It bounced like a waterbed as we walked on it. The permafrost below had thawed. (When you are in the presence of geologists, permafrost "thaws," never "melts"; no one else seems much bothered by the distinction.) It was this mossy, boggy Donut hole that the scientists wanted to investigate.

The difficulty in taking core samples from a bog is not so much acquiring the samples as it is keeping them in their vertical order. The stratigraphy, the top-to-bottom order, is the important part. As a general geological rule, the deeper the material, the older it is. History is written from the bottom up. But bog muck is wont to rearrange itself, given the chance. Scientists have developed a number of techniques to fight this tendency.

The first tool the USGS crew used was the frozen finger. It was a seven-foot pipe with a conical tip. The way it works is the same as how the kid in *A Christmas Story* got his tongue stuck to a freezing flagpole. The scientists shoved the pipe cone-first down into the muck, stuck a funnel in the open end, then poured it full of a slurry of dried ice and denatured alcohol. ("Anybody want a vodka tonic?" offered biogeochemist Mark Waldrop.) The bog was thus made to lick the pole. When the scientists hauled the finger up from the depths, it had acquired a rind of frozen bog muck. Most of the scientists retreated with this bounty to the Donut and began chipping the frozen muck from the pole, centimeter by centimeter. They dropped these segments into plastic bags, each bag carefully marked to indicate the stratigraphic position of its contents. These went into a cooler.

Miriam Jones and Mark Waldrop remained behind, probing the Donut hole with a second implement, called a Russian corer. The corer was a long pole with a T-shaped handle, tipped with a rounded spearhead that was conical along one side and flat on the other. By plunging the corer to the desired depth, giving the handle a half-twist in one direction, then a half-twist in the other, the scientists could pull plugs of peat fifty centimeters (about twenty inches) long—intact, stratigraphically sound—from a dozen feet below. The plugs smelled of mud and moss. Jones and Waldrop slid them into sections of PVC pipe, cut lengthwise, which they then bound with Saran wrap.

These plugs were full of proxies. Most obvious were the remains of the plants that once lived there, the macrofossils. (This only means they were visible to the eye; the remains in question were neither very large nor mineralized.) Macrofossils tend not to travel very far, Jones said, making them a reliable record of local conditions. A thick layer of peat moss indicates a bog or fen. Layers of charcoal suggest nearby fires. Other proxies were more arcane. Back at her lab in Virginia, Jones would use the relative concentration of carbon isotopes in the peat to precisely date the layers. The bog was 11,700 years old, she told me later, exactly the same age as the Holocene epoch. She would use oxygen isotopes to determine whether the rain that fell on the bog came straight from the ocean or traveled overland first—a way of studying the region's ancient climatic patterns. Finally, by comparing the amount of carbon contained in peat from the unthawed Donut with the amount contained in peat from the thawed Donut hole, she could estimate how much carbon had been released into the atmosphere, and perhaps get some hint of what might be in store as permafrost thaws around the world.

The plugs of peat moss also held a history of the area's Holocene forests, written in pollen. In order for seed plants to sexually reproduce, pollen must find its way from the male cone (of a conifer) or the anther (of a flowering plant) to a female cone or to the stigma of a flower, respectively. Some plants rely on animals to broker this exchange. Others cast their fate to the wind. Their pollen appears to us as a yellowish dust. This annual ejaculation is messy, wasteful, and allergenic. It also provides the most widely available record of ancient forests.

Pollen is nearly indestructible. Pollen grains buried 10,000 years ago might appear as fresh as the grains that plants produced last year. In order to extract pollen from the detritus that surrounds it, palynologists bathe the material first in an acid, then in a caustic. This dissolves away plants, charcoal, and rock. Only pollen is left. Next the palynologist must identify which pollen is which, grain by tedious grain. The pollen of every genus looks different. Some looks like a spiky pool toy, some like pasta, some like a Mickey Mouse head. After counting the pollen, the palynologist plots the relative frequency of each genus through time. (Many pollens are identifiable only to the level of genus, not to the species.) Palynological records often take the form of a series of connected bar graphs, showing a dozen or more types of pollen arrayed along one axis, with time along the other. Over the millennia, different plants become more or less frequent. In the same way that peat moss suggests a bog or fen, one group of plants on a palynological graph might suggest one physical setting, and another group of plants, something quite different. In the wax and wane of pollen, a palynologist can even watch the climate change.

A hole like the one the scientists poked in the Donut can say only so much by itself, of course. "We're just looking at one tiny hole in the ground," Jones told me. "But then you hope you can piece it together with other records." All across the state, scientists have poked similar holes in bogs and lakes and other places where useful proxies tend to accumulate. From these scattered hints they have stitched together the story of Alaska's post-glacial forests.

Where once grew sequoias, now grow spruce. Alaska's southern coast is fringed with Sitka spruce to the east, white spruce to the west, and in the middle a hybrid of the two, called Lutz spruce. In the state's interior, white spruce dominates the uplands, and black spruce, nearly everywhere else.

There are parts of Alaska where the black spruces seem to go on forever. They grow in monocultures that would make a farmer jealous—thick and dark and scraggled. They look like vertical dust bunnies, with branches distributed unevenly on their spindly trunks, gathered about their midsections or furled toward the top. They have prickly needles, dangling boogers of old-man's-beard lichen, and flaky bark that oozes little pustules of sap. Black spruce suggests its setting as strongly as a saguaro cactus suggests the desert. You find it in places like the Donut—on wet, low ground overlying permafrost, in places that are buggy, mossy, hard-traveled. Like a city pigeon, black spruce is indifferent to its lodgings, content to grow on acidic, impoverished soil, happy to live on scraps. It is just one of a handful of trees capable of pushing the water from its cells into the spaces between and so escape the fatal damage that being frozen solid usually does to living things. In good conditions, black spruce can get to be a big tree, but conditions are almost never good. Black spruce is a badlands tree.

The spruce forest is a scratchy sweater draped across the shoulders of the continent, stretching from Alaska to Newfoundland, down into the midwestern states and in patches as far south as New Jersey. This North American forest is, in turn, just part of a still bigger forest, a great boreal band that cuts across Russia and Scandinavia. The boreal contains a third of the world's forested area. I have spent much time flying over Alaska in small planes, and I once drove from Montana to Anchorage, where I grew up and where my parents still live. Seen this way, the boreal feels endless. What giant sequoias achieve by individual stature the boreal forest achieves by collective effect. It gives the strong impression of permanence. But the fossil record suggests that the boreal is not as steadfast as it now appears.

As Alaska warmed in the late Pleistocene and early Holocene, the steppe-tundra that covered the central part of the state thickened with shrubs, then gave way to a forest of deciduous trees: cottonwood, aspen, alder, and birch. About ten thousand years ago, this forest was replaced by one dominated by white spruce. Finally, about five thousand years ago, black spruce rose to dominance. Alaska's forest as we know it is only a little older than the Great Pyramid of Giza. (The dates and sequence are different in some parts of the state—Alaska is huge and varied—so this story should be taken only in a general way.) To an ecologist, these three forests are proxies for different things: different types of animals, different seasonal rhythms, different flows of carbon and nutrients. They suggest entirely different landscapes. The question is why one forest changed to another.

The first and simplest explanation is that these shifts in forest type reflect changing physical conditions. In modern Alaska, aspen and cottonwood tend to grow in drier areas, on recently disturbed ground, or in places with thin mineral soils. They can be found on river washes and on the moraines of retreating glaciers. Black spruce, as mentioned before, appears in wet, mossy places with thick organic soils, often over permafrost. White spruce's typical habitat lies somewhere between these extremes. Today, it tends to dominate the uplands. These rough impressions match what proxies tell us about Alaska's climatic history. The early Holocene in Alaska, when deciduous trees were dominant, was a little warmer and drier than today. Soils were poorly developed. In the middle Holocene, when white spruce was dominant, Alaska was still warmer than today but wetter than in the early Holocene. The state's climate continued to cool through the middle and late Holocene. Organic soils had developed. Peatlands expanded, and permafrost spread. Black spruce rose to dominance.

But physical conditions alone can't fully explain the shifts from one forest to another. All three types of forest are still widely distributed across Alaska. It is their relative dominance that has shifted. In other words, maybe what changed wasn't the various species' ability to survive, but rather their ability to compete. Recall Evelyn Hutchinson's fundamental niche: a hypervolume containing all the sets of conditions under which a species might survive, grow, and reproduce. Although I failed to mention it before, Hutchinson's niche had a second part—the "realized niche." It is another theoretical, hypervoluminous space, this one containing all the sets of conditions that a species will utilize after accounting for its interactions with other species. The realized niche would tend to be smaller than the fundamental niche, Hutchinson wrote, because in many otherwise survivable sets of conditions, the species would not be able to compete, or it would be eaten, or it would not have the other species it depends on.

At the broadest scale, evolutionary biologist Town Peterson told me, the arrangement of species reflects physical conditions. But on a local scale, interactions between species might matter more. The U.S. Geological Survey range maps of white spruce and black spruce, for example, are almost identical. Within those ranges, though, the two spruce species tend to occupy different types of places. While the spruces might have roughly similar fundamental niches, they have different realized niches—different sets of conditions in which they are capable of competing. Maybe this explains the sudden shifts between Alaska's three distinct Holocene forests. A small degree of change in physical conditions tipped the odds of competition. The places where the seeds of each species tended to succeed or fail shifted, and tree by tree, the landscape was transformed.

But there is a third possible reason for the shifting arrangement of Alaska's Holocene forests. Perhaps deciduous trees dominated in the early Holocene not because spruces couldn't bear the conditions or couldn't compete, but simply because they hadn't yet arrived.

One day in the fall of 1895, Clement Reid flushed a flock of rooks from an oak-fringed field. "On driving the birds away, and walking to the middle of the field, I found hundreds of empty acorn-husks, and a number of half-eaten pecked acorns," he wrote. Several of the acorns were still intact. One, he wrote, had even been buried "by a single peck deep into the soft soil of a mole-hill." This was exactly what he'd been looking for.

Reid was an English geologist. He was puzzled by what seemed to be a strange contradiction. He had made a careful study of the movements of the oaks of Britain. Acorns don't fall far from the tree, and it takes a long time for oak seedlings to grow and produce their own acorns. Modern oaks, he'd concluded, were slow. The fossil record showed that the oaks had spent the last Pleistocene glaciation far to the south, in Spain and France. For the trees to return to England, he wrote, "would take something like a million years." But the fossils also showed that oaks had made it back to England only a few thousand years after the end of the ice age. Fossil oaks were fast. It was a paradox.

The answer to this paradox, Reid thought, was birds. Maybe rooks. He suggested that the oaks that had colonized the British Isles had arrived not as a single wave of advancing trees, the way people usually imagined the migration of forests, but instead in a sort of relay race. Most acorns fall close to their parent trees, but on occasion a bird will carry an acorn far away from the main forest. Once in a while, one of those birds will drop its acorn intact onto a patch of fertile ground—maybe a molehill. And once in a while, one of these acorns will sprout and grow up to be a tree that produces its own acorns, which, on occasion, a bird will carry far away.

As in northern Europe, the fossil record in North America seemed to show rapid movement of many tree species at the end of the last ice age. In a 1986 paper, J. C. Ritchie and Glen MacDonald estimated that as the ice sheet retreated at the end of the Pleistocene, white spruces had raced north across Canada at more than a mile per year. This could perhaps be explained, they wrote, by strong northwesterly-trending winds off the retreating ice sheet.

In the late 1990s a group of researchers revisited the question of Britain's oaks and provided a mathematical framework for Reid's paradox. Assuming some small number of long-range plantings by birds, storms, or other unusual events, forests could move fast, they wrote, maybe even infinitely fast, at least from a statistical point of view—"a maximum rate [of travel] might not exist." Reid's ice age oaks might have been capable of going even faster if the plodding rate of climate change hadn't held them back.

In the context of modern climate change, this was good news. It meant that trees had the potential to keep pace and arrive in those places made newly suitable. It meant it was possible to imagine the world's forests maintaining roughly their present form, just shifted a bit poleward and upslope.

But Jim Clark, the biologist who led the group of researchers, told me that the team soon realized that its mathematical model poorly reflected the physical traits of actual trees and their seeds. The more tree-specific information they added to their model, the slower the simulated trees became.

Many of the random acts of long-range tree planting supposedly driving this rapid movement made little practical sense. Acorns and birds, sure, Clark said, but hickory nuts? "There is really a small number of animals that deal with hickory nuts," he said. A hickory nut is more trouble than it's worth. "If you had a rock, you could bust it open," he said. "But it takes so much work to do, and you'd smash the things inside. It's kind of a mess." Such problems abound.

In 1937, a Swedish botanist named Eric Hultén offered another solution to Reid's paradox. Imagine, Hultén wrote, "a large area of suitable soil cleared from all vegetation and allowed to develop under equable and favorable conditions without competition from other plants." In this paradisiacal patch, a plant would suffer no shortage of water or nutrients or space. It would grow and reproduce as well and as fast as it possibly could. Now imagine setting multiple species of plants in the center of this patch. They would grow and spread out from this center, Hultén wrote, "but it could hardly be expected that all plants would spread with the same rapidity." Some species would go fast, others slow, and before long the patch would take on the appearance of a target: concentric circles of plants, with the slowest species near the bull's-eye and the fastest near the outside. This hypothetical scenario, he pointed out, is not so different from what happened at the end of the Pleistocene, when North America was "laid bare by the retreating ice."

Hultén had spent the previous decades wandering Scandinavia, the Russian Far East, and finally western Alaska's Aleutian archipelago, the island chain slung like a hammock between North America and Asia. On this last trip, in 1932, he traveled with an American assistant, Walter Eyerdam. For most of the year they worked their way back and forth between the windswept islands, hitching rides with the Coast Guard and trading vessels. Eyerdam later recalled that on a climb up Ballyhoo Mountain, outside Dutch Harbor, Hultén and a German sea *Kapitän* debated the relative importance to society of a learned botanist and a military man. "Each, of course, could see only his own profession as the more important," Eyerdam wrote, noting, "Today, at least in the larger countries, nearly all governments relegate all branches of science to the support of militarism."

Hultén was interested in the ranges of the plants of the boreal. As he traveled across Scandinavia, the Soviet Union, Alaska, and Canada, he made maps of where the various plant species appeared and where they disappeared. When he combined these species-range maps, stacking them one on top of another, what he found was a series of targets—concentric rings of plants spreading out from bull's-eye centers. It was just like his hypothetical garden, except that in real life the targets were oval, not circular, pinched by the drop in temperature in the north and the increase in aridity to the south. The plants had spread out from these centers, Hultén thought. The bull's-eyes marked the places where they'd waited out the Wisconsin glaciation.

Hultén's manner of inference was innovative, but the real surprise was the location of these bull's-eyes. "It is often presumed that the ice, as it proceeded, 'chased' or forced the plants south," Hultén wrote. "In my opinion, this was not so, at any rate not in most cases." The bull's-eyes he'd found were in northeastern Siberia, in Manchuria, in the Altai Mountains of central Asia, in northern Japan, and in the Yukon, where it had been too dry for glaciers to form. They lay just beyond the edge of the ice. Hultén called these bull's-eyes "refugia"—places of refuge where species were "left in possession of a small part of their earlier area, and where they were able to survive the severe conditions of the maximum glaciation."

Scientists today think that many plants did weather the last Pleistocene glaciation in northern refugia, although they continue to debate exactly where these refugia were and what role their residents may have had in repopulating the region after the retreat of the ice. There is a dearth of physical evidence to back up Hultén's inferences. The populations in northern refugia were likely small, leaving little trace. Among the few macrofossils scientists have found to suggest the presence of spruce in glacial-era Alaska and northwestern Canada are a stump, a handful of needles, and a few cone scales, some of them discovered in the gut of a mummified horse. The palynological evidence is scant, too.

In recent decades scientists have begun using another kind of inference. When a small population of organisms is separated from others of its species, it tends to experience genetic drift, where certain genetic variants become predominant or disappear from the population. Eventually, this population will gain a measurable genetic difference from the larger population. This difference can persist for some time, even after the two populations reunite. By measuring this difference, scientists have shown that both black and white spruces survived the Wisconsin glaciation in Alaska, joined only much later by their kin from the south. Looking back on the race, it seemed as though the trees had made amazing time, when really we'd imagined the starting line in the wrong place.

Still, most of the scientists I talked with thought that Alaska's three distinct Holocene forests were shaped more by the effects of climate and competition, and less by the various species' ability to reach suitable places. But that may not always be true. In a 2004 paper, ecologist Jens-Christian Svenning and botanist Flemming Skov modeled the climatic niches of fifty-five species of European trees. (I'll explain the process in a later chapter.) Thirty-six of those trees, they found, occupied less than half of the space on the continent that was climatically suitable to them. "Consequently," Svenning and Skov wrote, "we expect European tree species to show only limited tracking of near-future climate changes."

Even supposing that the apparently rapid rates of tree migration we see in the fossil record truly reflect what trees are capable of, Svenning told me, those rates might not be a good guide to what we can expect for most trees in the future. The tree species that appear to have traveled the fastest in the early Holocene were those expanding over land scraped bare by Pleistocene ice. "There was very little competition," he said. The species further back in the pack, though, had to not only arrive but also compete with those that had already established themselves. "A small minority moved very fast," Svenning said, "but most moved very slow." The same will be true for trees today, he said. While some might find open lanes, many will be stuck in heavy traffic.

In theory, at least, it should be easiest to see and understand the movement of a forest at the tree line, that place where trees arrived and went no farther. In recent decades scientists around the world have observed the tree line shifting poleward and upslope. But in many places, what they've observed further undermines any idea of species responding to climate change in a predictable, orderly way.

In the summer of 2017 I joined a group of researchers on a trip to the tree line in the White Mountains of eastern California and western Nevada. To the west, across Owens Valley, rose the craggy jawline of the Sierra Nevada. To the east lay basin and range, longitudinal rows of mountains with circular alfalfa fields in the flats between. A little to the northeast was a light so bright it looked like a second sun. It was a solar plant, in which a circle of banked mirrors heated a tank of molten salt, in turn driving a steam generator. The beams from the mirrors were known to cause passing birds to spontaneously combust. The scientists I was with called it the Eye of Sauron.

The White Mountains have a backbone of dolomite, tan rock that goes nearly white in the midday sun. Lying in the rain shadow of the Sierra Nevada, they are dry and inhospitable. There, at about twelve thousand feet of elevation, most of the plants were of a dwarven mien, huddled between the dolomite blocks. The botanists peered at them through hand lenses, hunched over, squint-eyed, cyclopean—is that *Elymus multisetus* or *Elymus elymoides*? *Festuca baffinensis* or *Festuca minutiflora*? I'd given up. They were all just a bunch of little grasses to me. Instead, I focused my taxonomic attention on the trees, of which there were a manageable two varieties: Great Basin bristlecone pine (*Pinus longaeva*) and limber pine (*Pinus flexilis*).

The bristlecones are the world's oldest single-stemmed trees, sometimes living more than five thousand years. One of the oldest known bristlecones was at least 4,844 years old when it was felled in the mid-1960s by a U.S. Forest Service crew working at the behest of a graduate student who hoped to determine its age. The bristlecones are twisted like croissants, kneaded by wind and snow, baked by the Sun. Sometimes in dry spells they will allow part of their cambium to die, an austerity measure. Over the centuries, the bark overlying the dead cambium falls away, leaving the trees in advanced states of undress, naked wood weathered to a high luster. Often only a Speedo strip of bark remains, covering a last bit of living flesh that winds its way up to a single tuft of needles. People are impressed by the bristlecones. Alaskan folk singer Hobo Jim sang a ballad in their honor. The limber pines, meanwhile, reach only one or two thousand years of age and have a less tortured physicality, seeming to bear their centuries more lightly. Superlatives are relative, so the limber pines are afforded proportionately less respect. There are no songs about limber pines.

The scientists were surveilling the mountain's plants, noting which species were growing along tape lines they'd laid at regular altitudinal intervals. By positioning the tapes in the exact same places year after year, they could roughly track the floral traffic. When I could avoid the botanists trying to enlist my help laying tape, I attached myself to Brian Smithers, then a doctoral candidate at the University of California, Davis, and the organizer of the expedition. (It was part of a worldwide project called GLORIA: GLobal Observation Research Initiative in Alpine environments.) Like me, Smithers mostly ignored the various dwarf-grasses and thimble-flowers, focusing instead on the trees. Smithers competes in triathlons and brews his own beer and has an effortless beard. His peers had given him a nickname that was a prurient pun on the limber pine's Latin name.

During the Wisconsin glaciation, he told me, the two trees lived in Nevada, at elevations now favored by alfalfa. They arrived here on the mountain sometime in the early Holocene. The living trees at this tree line are perhaps of the second or third generation that sprouted after the Pleistocene cold. In the warm period at the beginning of the Holocene, the trees made it even farther up the mountain. You can find groves of weathered snags another five hundred feet higher. Now, as the climate warms, you might expect the trees to push back into that elevation. Indeed, the day before, while wandering alone through the groves, I'd found several small bristlecones hundreds of feet upslope of the groves of adult trees. I mentioned this to Smithers, excited about the tidy narrative: The trees used to live higher up, but then, as the planet wobbled its way through the Milankovitch cycles in the latter part of the Holocene, it got colder and they died. Now it's getting warmer again, and they're spreading back up the mountain.

But Smithers squashed my story. The situation was messier then I'd imagined. "You have multiple layers of advance and retreat, advance and retreat," he said. To figure out which trees lived and died when, he said, "you'd have to core a lot of trees." The groves of dead trees could have been killed not by cold but by drought. The young trees I'd seen might have had nothing to do with modern warming. They could have sprouted hundreds of years ago, during the Medieval Warm Period (a stretch of slightly warmer climate during the Middle Ages). Youth, too, is relative.

Smithers has found that the tree line in the White Mountains is indeed moving, but not in the straightforward way I'd envisioned. While bristlecones currently dominate the tree line, limber pine seedlings seem to be winning the race upslope. The switch in position is hard to account for. The bristlecones have light, wind-blown seeds, while the limber pines depend on the Clark's nutcracker, a jaylike bird, to carry them around. Maybe the bristlecones arrived here first after the Pleistocene and kept the limber pines from claiming the tree line. Or perhaps the limber pines arrived first thousands of years ago and gave shelter to bristlecone seedlings, which eventually grew up to supplant them; now, millennia later, no trace of these nursery trees remains. Or it could be that the wind-dispersed bristlecone seeds were more likely to find a favorable patch of ground when the rate of climate change was relatively slow, but now that the rate of climate change is fast, the bird-dispersed limber pine seeds are capable of traveling farther into newly available terrain. Perhaps the apparent switch taking place today will prove ephemeral, Smithers said. "Maybe a ten-year cold snap kills all the limber pines, and the bristlecone pine turns out to be a little hardier." He couldn't say for sure. We were standing too close to make out the whole story.

What was more certain, Smithers said, was that over the last fifty years, the tree line in the White Mountains has shifted upslope by about one hundred meters, or two meters (six and a half feet) per year. So far neither bristlecone nor limber pine is winning the bigger race, the race to keep up with the changing climate.

The world continues to wobble through its Milankovitch cycles, and by now it should be cooling again. As John and Katherine Palmer Imbrie wrote in *Ice Ages*, published in 1979, "Most scientists who have examined the evidence agree that the world will experience another age of ice." The question was only when. Some scientists, the Imbries wrote, "believe that an ice age is already on its way—due within the next few centuries, according to one extreme view."

But the world is not cooling. Scientists have known for more than a century that carbon dioxide, a product of decomposition and combustion, and an essential ingredient of photosynthesis, is an efficient insulator, and that even a slight rise of its concentration in the atmosphere would warm the planet. For more than fifty years, scientists have warned of what might happen if humans continued to burn fossil fuels. So far, we have mostly ignored these warnings. As E. C. Pielou wrote in *After the Ice Age*, "the measured responses of biosphere to climate, and of climate to astronomical controls have, for the foreseeable future, come to an end."

In 2000, atmospheric chemist Paul Crutzen and biologist Eugene Stoermer proposed that stratigraphers delineate the present and recent past as a new geological epoch. They offered a name, the Anthropocene—the Age of Man. Critics have called the Anthropocene epoch geologically arbitrary, and proponents have argued at length over when it began. But the concept has caught on, serving its intended social purpose, at least. It is a name to remind us that we've altered the trajectory of the entire world—land, ocean, and atmosphere alike. As I write this, in the summer of 2019, sensors on Mauna Loa, on the Big Island of Hawaii, have measured the atmospheric concentration of carbon dioxide at 414 parts per million, up from about 280 parts per million just before the Industrial Revolution, and less than 200 parts per million during the last Pleistocene glaciation. The last time the concentration was this high was between 3 and 4 million years ago, in the mid-Pliocene. Recently a group of scientists estimated that by the middle of this century, we will reach a carbon dioxide concentration not seen since the beginning of the Eocene epoch, 55 million years ago, when metasequoias grew in the Arctic.

The rate of change is faster than in all but the most violent periods of change in the past. Scientists sometimes refer to the changing climate in terms of "climate velocity"—that is, how fast you would need to travel over Earth's surface in order to maintain steady climatic conditions. In 2009, a group of researchers led by Scott Loarie projected that between 2000 and 2100, the average climate velocity on land under intermediate scenarios of warming would be about a quarter mile per year. This velocity depends on topography. The distance you have to travel north to maintain the same climate is much farther than if you were to travel upslope. In some flat regions, the velocity could be more than five miles per year. Over flat ground, even a tree species at full gallop might be left behind.

When the Wisconsin glaciation hit its zenith, some 21,000 years ago, the swollen ice sheets held so much water that the level of the sea fell by nearly four hundred feet. Between Alaska and Russia, in the space now filled by the Bering and the Beaufort Seas, a vast plain lay exposed. It has been called a "land bridge," but it was really more like a subcontinental extension of Asia, bounded to the west by the Verkhoyansk Mountains and to the east by the Mackenzie River. Most of this land was too dry to have supported an ice sheet. Some of Eric Hultén's overlapping ovals of plant ranges had their centers in this plain. "This landmass," he wrote, "which I shall hereinafter call Beringia, must have been a good refugium for the biota during the glacial period."

Scientists think that Beringia was indeed a place of refuge for many species, although they debate what it looked like. Much of the landmass is now deep underwater, leaving few clues. Some scientists I spoke with called it an endless, impassable swamp. Others described it to me as a colder version of the modern African savanna, filled with bison, mammoths, elk, and other large mammals—as fruitful, from a hunter-gatherer's perspective, as anywhere on Earth.

When exactly humans continued past Beringia and into North and South America remains a topic of debate; archaeologists' estimates range from as recently as 14,500 years ago to 32,000 years ago or even earlier. By even the most conservative estimates, though, people were in North America in time to witness the return of the forest.

A month after my trip to the Homer beach with Ed Berg, archaeologist Josh Reuther took me to a game ranch near the town of Delta Junction, in Alaska's interior. We watched as one of the ranch's owners and a client performed a modern post-hunt ritual.

"Hold on," said Scott Hollembaek, the owner—he wanted to fix the tongue. He was lying on the ground, dressed head to toe in twelve-ounce canvas, holding a digital camera. The client was kneeling behind a dead bison, which had its tongue out. Hollembaek hauled himself up, pulled on a pair of blue doctor's gloves, and carefully stuffed the bison's tongue back into its mouth. Retaking his prone position, he snapped one photo, then another. "Hold the rifle different, so not all the pictures look the same," he told the hunter. "And change your smile, too."

We were in a large clearing. The surrounding forest was mostly deciduous trees—birches and aspens—with scattered white spruces. It was open, with a grassy understory. This is how Alaska's forests looked in the early Holocene, 10,000 years ago, Reuther told me. Alaska was then full of large grazing animals, like the elk and bison that now populate Hollembaek's ranch. A few days earlier, I'd visited another archaeologist, Ben Potter, at the University of Alaska Fairbanks. He told me that the Beringian hunter-gatherers of the early Holocene had a lifestyle unlike that of any modern hunter-gatherers. They had an intricate material culture, he said, a proxy for people who had time to spare. They had no food caches, suggesting a life of plenty. Their numbers were few, and they had no set territories, instead wandering as needed in pursuit of the best hunting and the most abundant gathering. They lived as if the world were boundless.

Hollembaek backed his tractor and trailer up to the bison, wrapped a chain around one of its hind legs, then used a winch attached to a chainsaw motor to drag it up onto the trailer. He drove off with the hunter, leaving Reuther and me behind in the clearing. Reuther led the way across the field and up a steep south-facing slope, warmed by the midday sun. There was the sweet smell of sage. Here on this slope was a small remnant of what much of Alaska looked like during the Wisconsin glaciation, Reuther said: a mix of sagebrush, grasses, and sedges—what scientists call the "mammoth steppe." A few south-facing hillsides like this one are all that's left of an ecosystem that once covered much of the state.

We reached the top of the rise. At the crest was a fire pit, stacked logs, an empty beer can. Hollembaek's clients sometimes sat up there and watched the animals, Reuther said. It was a perfect place for spotting game, a high prominence on an otherwise flat floodplain. "You can see two hundred seventy degrees up here," Reuther said, looking around. That was how archaeologists Chuck Holmes and Randy Tedor found the place. They had been poring over topographical maps, trying to imagine where ancient hunters might have hung out, when they noticed the promontory. They called Hollembaek and asked if they might have a look around. Hollembaek agreed. Reuther and his colleagues began excavations a few years later. "Sure as shit," Reuther said—just a few feet up the hill from the modern fire pit, he found another, 13,500 years old.

People used the hillside for thousands of years. "They camped up here, hunted down there," Reuther said, gesturing to the clearing where Hollembaek's client shot the bison. Then, around 5,000 years ago, the forest of white spruce and aspen and birch began to give way to black spruce and moss. The land grew wetter, boggier, harder to travel. By then, the early Beringians had spread out and multiplied. Most of the big mammals were gone. Some, like horse and mammoth, seem to have been in decline before people arrived. Others may have declined because of people. The paleoecologist Paul Martin argued that even a small number of humans could have killed off slow-breeding mastodons, bears, horses, and other large mammals without meaning to, without even noticing. As Aldo Leopold memorably put it in *A Sand County Almanac*, "The Cro-Magnon who slew the last mammoth thought only of steaks."

When elk and bison disappeared from Alaska, the people turned their attention to salmon and moose, Ben Potter told me. They grew more territorial and began to make food caches—a sign, he said, that they'd come to know seasonal scarcity. Around the same time that black spruce rose to dominance, hunters stopped camping on Hollembaek's hill.

Reuther and I stood there on the hill, looking west to the shining, braided form of the Tanana River. Just below us was the tiny patch of the mammoth steppe, a relict of the Pleistocene. Beyond that was the forest of birches and aspens and a few white spruces, a small piece of the forest of the early Holocene. Beyond that was spruce forest, now extending for hundreds of miles in every direction.

I wondered what the ancient people who lived there had thought of their changing world. Did they notice the spruce seedlings coming up, or that, over the years, the animals they hunted seemed to grow fewer and fewer? Or maybe it was subtler than that. At present, the climate is changing far faster than it ever has in the past, but even now, it can be hard for a person living through this transformation to judge what is normal, what is abnormal, and what is wholly new. Maybe back then it was just stories told by the elders of how the world used to be different, conditions better, game more easily had. The baselines shift so subtly as to elude perception. I wondered whether they worried about what was coming.

HOW MONTEREY PINE BECAME RADIATA AND OTHER STORIES

For most of their history, people were focused on where things already lived. In order to hunt and gather, hunter-gatherers needed to know the habits and haunts of animals and the places and seasons of edible and useful plants. What was important to the hunter-gatherer was the world as it was presently arranged.

Then, toward the end of the Pleistocene, people began to turn their focus. A few millennia after the first hunter-gatherers flooded out of Beringia into the Americas, people in the fertile land surrounding the Tigris and the Euphrates Rivers of modern-day Iraq became farmer-gardeners. They gathered useful species around them and settled into villages. Unlike the hunter-gatherers, these farmer-gardeners were interested in where things might live, if given the chance. A certain species might not thrive in a certain place now, but what would happen if you pulled out the competing plants and dug trenches for irrigation and fended off herbivores; might it survive then? The farmer-gardeners' position was more open-ended, more grandiose. They might look at a forest and imagine a field. They might see a plant and wonder what would happen if you planted it somewhere new. The farmer-gardener believed the world could be changed to fit.

I admit that this story as I've laid it out feels a bit just so. The hunter-gatherer and the farmer-gardener are archetypes, but really there is no hard divide between them. Many hunter-gatherers used fire and other tools to shape the land long before the invention of agriculture, and many farmer-gardeners continued to hunt and gather long after they started planting. But it is fair to say that, through the Holocene, people around the world trended toward the habits of the farmer-gardener. By the time Christopher Columbus spotted the islands of the Caribbean in 1492, the majority of the world's people leaned farmer-gardener.

Jared Diamond wrote in *Guns, Germs, and Steel* that early anthropologists tended to sum up the life of the ancient hunter-gatherer with the Hobbesian nugget "nasty, brutish, and short." These anthropologists were themselves committed farmer-gardeners. They believed their approach to be the better one. But in recent decades anthropologists have given this conclusion a second look. The hunter-gatherers actually had it pretty good, they say. In *Sapiens*, Yuval Noah Harari wrote that hunter-gatherers spent less time laboring to feed themselves than the farmer-gardeners who followed. They had better teeth, longer lives, and "physical dexterity that people today are unable to achieve even after years of practicing yoga or t'ai chi."

Less work, better teeth, and rockin' bods are among the things modern people most want, so the ancient transition to farmer-gardener could indeed be read as a mistake. But however Hobbesian the lives of individual farmer-gardeners, their focus on where things might grow has allowed our species amazing success. We billow toward eight billion.

This success has come at a heavy cost to the rest of the world's life-forms. The hunter-gatherers were hard on the world's large fauna, but they were relatively easy on everything else. The farmer-gardeners are hard on nearly everything. The very goal of agriculture is to favor one thing over another. Humanity has put its weight behind the family Poaceae, the grasses. It is largely in the service of barley, wheat, rice, and maize that we've felled the world's forests. The farmer-gardener, who looks at a forest and sees a good place for a farm, is not generally concerned with where a given plant or animal came from. She is instead interested in how useful the organism might be, and whether with a little bit of help it might survive in new places. To a farmer-gardener, the fact that different things grew in different parts of the world is not just a quirk of biogeography but a problem to be solved.

One such problem: Ancient Egypt had no frankincense trees, the source of fragrant pitch, so in 1500 B.C., Queen Hatshepsut sent ships to the land of Punt to bring some back. (It's not clear where on the modern globe Punt is, although most evidence points toward the Horn of Africa.) The journey was the first recorded instance of an expedition specifically in pursuit of live plants. It was a success. A mural at the Deir el-Bahari shows the trees with their roots bound in baskets. Teams of men used ropes and poles to carry them up gangways into the ships, earning the trees the unusual distinction of making not one but two journeys in their lives, the second completely uncoupled from the geographic path blazed by their ancestors.

The Romans took a similar interest in improving their surroundings. They planted stone pines in Greece, Turkey, Israel, and North Africa, and olives, wheat, and barley everywhere these crops would grow. To the British Isles, they brought plane trees, chestnuts, elms, pears, cherries, figs, quinces, onions, cabbages, leeks, radishes, and opium poppies. The historian Tacitus reported that, sadly, the islands could not sustain wine grapes.

It was Europeans' culinary tastes that eventually led them to open the widest of botanical frontiers. As Carolyn Fry wrote in *The Plant Hunters*, the fall of Constantinople to the Turks in 1453 cut off overland routes from Europe to the spice markets of Asia. Christopher Columbus was among the explorers who went looking for another way to the Indies. The conquistadors who followed plundered the civilizations of the New World in search of gold and jewels, but the maize, potatoes, and beans they carried home were as valuable as any mineral. "Man," wrote Alexander von Humboldt, "being restless and industrious, traveled in all the earth's regions and thereby forced a certain number of plants to live under many climates and in many altitudes." The farmer-gardeners planted old species in new places and new species in old places and asked, again and again, Will it grow?

Over the millennia, these explorations revealed something surprising. As the farmer-gardeners tested which things would grow in which places, they found that often the present location of a species revealed little about where else that species might thrive. From a modern perspective, this seems obvious. We're surrounded by species that live in far different situations from those they knew in the wild. But before farmer-gardeners had carried a significant proportion of the world's species to new places, there would have been no way to know if this might be true. This realization meant that any divine arrangement—each beast to its place and each plant to its patch—was at least flexible. It meant that rarity itself was not a sign that a species was somehow deficient. It meant that sometimes rare species were simply unlucky.

David Douglas was born in 1799 in Scone, Scotland. The second son of a stonemason, he was interested in natural history from an early age. He took an apprenticeship at the Scone Palace gardens, then at the Glasgow Botanic Gardens, where he worked alongside William Hooker, one of the eminent botanists of the day. When a member of the Royal Horticultural Society came looking for someone to make a society-sponsored botanizing trip to North America, Hooker suggested Douglas.

Europe's age of exploration was in full swing. Ships roved the sea and men stalked the land, searching everywhere for what was new and strange and useful. The herb gardens of apothecaries and physicians grew into botanical gardens and arboretums and greenhouses. Botanists tested plants from around the world, probing what they could be good for and where they might be coaxed to grow. Wealthy landowners competed to stock their gardens with the most novel and beautiful plants. Behind this botanical confluence of science, industry, and hobby were the plant hunters themselves, who traveled the far corners of the world, collecting samples, seeds, and live plants. Some, like Alexander von Humboldt and Joseph Banks, were amateurs, men of means who paid their own way. Others, like David Douglas, were professionals.

Douglas was twenty-four when he left Liverpool aboard the *Ann Maria,* bound for New York. After a two month's sail (his journal here records each day's variety of fog: "rather foggy," "thick fog," "very unpleasant fogs"), he arrived at Staten Island in August 1823. He spent the next four months traveling around New England and Canada.

During a short trip to the Canadian side of the Detroit River, Douglas had a first brush with the misfortune that would eventually catch him. He was up an oak tree, collecting acorns and leaves. It was a warm day, and before he climbed the tree he had taken off his coat. "I had not been above five minutes up," he wrote, "when to my surprise the man whom I hired as guide and assistant took up my coat and made off as fast as he could run with it." Douglas scampered back down the tree but couldn't catch the thief. Along with the coat, he lost nineteen dollars, the receipts for the rest of his money, his field guide, and his vasculum (a box used by botanists to carry samples collected in the field), along with the samples therein. He had to hire a man (on credit, presumably) to drive him back to his lodgings, because the horse he'd borrowed understood only French.

Still, the trip was a success. By the time Douglas arrived back in England, on New Year's Day 1824, he'd collected samples and seeds from dozens of plants, including nineteen species of oak. Within months of his return home, the Royal Horticultural Society enlisted him for another trip, this time to the Pacific Northwest. Douglas wandered the region for the next four years. In letters to colleagues, he told of his amazing discoveries.

But this life of exploration was a hard one. Douglas's journal of his second expedition is almost farcical. He nearly froze to death so often that it begins to feel repetitive. He suffered heat, thirst, and loneliness. Fleas and mosquitoes plagued him. He stabbed himself in the knee with a rusty nail and was laid up for ten days. Once, he was tossed from his horse into a river and lost his knapsack and notebook. Another time he fell down a gully while chasing a deer and lay unconscious for five hours. He got so hungry he ate lichen cakes. Horses died out from under him, and sometimes he ate them. Often as not, he lost the samples he'd collected. One night, he was visited by a pack of enormous rats, which, he wrote, "devoured every particle of seed I had collected, eat [*sic*] clean through a bundle of dried plants, and carried off my soap-brush and razor!"

When Douglas arrived back in England, he found that his stories proved popular in high society. "Flattered by their attention," Hooker later wrote, "and by the notoriety of his botanical discoveries, . . . he seemed for a time as if he had attained the summit of his ambition." But soon Douglas grew restless, Hooker recalled, "so that his best friends could not but wish, as he himself did, that he were again occupied in the honourable task of exploring North-west America."

There is scant record of Douglas's third trip to North America. What is known comes mostly from the letters he sent to his friends, especially to William Hooker. He found no passage to the Columbia River, so he was forced to spend several months around Monterey Bay, eighty-five miles south of San Francisco, where there was a Catholic mission.

There he found a tree that people would later call by the common name Monterey pine. Douglas named it according to Linnaeus's binomial system: *Pinus insignis*, Latin for "remarkable pine." It is unclear why he called it this. Compared with many of the other trees he had encountered—the sugar pine and grand fir and Sitka spruce—this tree was not especially impressive. "You will begin to think that I manufacture Pines at my pleasure," Douglas had joked in one letter to Hooker. Maybe he was just running out of good names.

This Monterey pine "displayed thoroughly unpretentious credentials," wrote Peter Lavery and Donald Mead in a history of the tree. The trees had deeply lined bark and whiskery, dark green needles in bunches of three. The pines were small and large, asymmetrical, imperfect, all hopelessly out of square. They grew intermixed with oaks and maples, madrones and cypresses, over a tangle of barberries, manzanitas, toyons, ferns, sedges, and many dozens of other plants.

Like the giant sequoia and the Florida torreya, the Monterey pine is a narrow endemic, native to just five small forests—three in California and two on islands off the coast of Baja. When Douglas found it, the tree grew across just a couple dozen square miles. In these areas the pine sometimes experienced frost but rarely snow. Ample summer fog dulled the thirst of summer. But this narrow set of conditions was rare and likely ephemeral, Lavery and Mead thought. By the long view, the tree "was being rapidly squeezed by changing climatic patterns, apparently towards extinction." But in meeting Douglas, the tree's luck suddenly improved.

Douglas wrote to Hooker from somewhere inland on the Columbia River on April 9, 1833. He sounded in good spirits, though he noted that he'd lost all vision in his right eye. He wrote next on May 6, 1834, from Oahu. He apologized for his long delay. After he'd sent his last letter, he wrote, he'd reached the Pacific coast "greatly broken down, having suffered no ordinary toil, and, on my arrival, was soon prostrated by fever." After his recovery, he'd capsized while boating on the Columbia and tumbled over a waterfall. "My canoe was dashed to atoms," he wrote. Knocked senseless, he was carried to shore by the current but lost nearly all his possessions, including his journal of the expedition and samples of the roughly four hundred species he'd collected. The lucky Monterey pine, *Pinus insignis,* was not among those lost.

Douglas told Hooker he'd sailed from the Columbia River in November 1833, arriving in Hawaii in January. He'd spent the following months exploring the volcanoes of the Big Island. "In this expedition I amassed a most splendid collection of plants," he wrote, "many, I do assure you, truly beautiful." He promised to give a full accounting later. He thanked Hooker for the letters the elder botanist had sent him during his journeys and added, "May God grant me a safe return to England."

On July 12, 1834, the wandering botanist was found dead on the flanks of Mauna Kea, at the bottom of a pit used to trap wild cattle. "His close [*sic*] are sadly torn and his body dreadfully mangled," wrote one Mrs. Lyman, the wife of a missionary, in her journal. Douglas's death at thirty-five remains a mystery. Did he wind up in the pit with a cow on top of him because of a careless stumble? It wouldn't have been the first time he'd planted his feet the wrong way. Was it because of his deteriorating eyesight? Or was it, as some speculated, that he'd gotten too friendly with the wife of his host, a cow hunter and former resident of the Australian penal colonies?

By the time of his death, Douglas had introduced hundreds of species of plants to the British Isles. Many of their progeny can now be found growing throughout the world. One of the most magnificent trees he'd discovered, *Pseudotsuga menziesii*, was given a common name in his honor: the Douglas-fir.

But one tree Douglas would not get to name was the pine of Monterey Bay. Thomas Coulter, an Irishman, had encountered the tree a few years before Douglas and had sent samples back to taxonomist David Don. Although it was Douglas's seeds that would travel the world, not Coulter's, in taxonomy naming rights go to the first person to describe a species. Don named the tree for the radiating lines on the scales of its cones. Douglas's remarkable pine became *Pinus radiata*.

On a cool morning in the antipodal spring of 2018, I was picked up at my hostel by Wink Sutton. "When I was a baby, my parents thought I was rather cute," he explained later. "They called me Wee Willy Winkie. It wasn't until I was about ten that I knew my name was really William." We were in Rotorua, a town in the central part of New Zealand's North Island that, owing to geothermal activity, smelled stubbornly of farts. Sutton, a retired forest geneticist and economist, was going to take me to see his museum. We'd met up the day before at the bus station. In a note with directions, he told me how to recognize him: "I am nearly eighty and walk with a small hunchback." We found each other without incident.

Now we drove northwest through sheep farms and tumbledown towns. I could pick out many familiar trees: cedars and oaks, sycamores and sweet gums, coast redwoods and Douglas-firs. They were trees of the Northern Hemisphere, the same species I see around my neighborhood in Oakland, California. But these were just ornaments. The background was dark green pines. They stretched over hillsides and fields, their trunks evenly spaced and parade-day straight, their boughs perfect spearhead points. They looked so at home I could almost imagine they had been there all along.

New Zealand was the last major landform on Earth to be inhabited by humans. When Polynesian voyagers arrived roughly 850 years ago, the islands were almost entirely forested. The Māori hunted the moas, the enormous flightless birds that stalked the forest, and lit fires to clear the land. By the time Captain James Cook arrived aboard the HMS *Endeavour*, in the late 1760s, the moas were gone and New Zealand's forests had shrunk by a third.

Joseph Banks was a naturalist aboard the *Endeavour*. As the ship circumnavigated the islands, Banks looked on with a farmer-gardener's eye. He noted the richness of New Zealand's soil and the "immense areas of woodland which were yet uncleared, but promised great returns to the people who would take the trouble of clearing them." He noted the favorable lay of the banks of the river that Cook called the Thames (now called the Waihou). The area was "indeed, in every respect, the most proper place we have yet seen for establishing a colony," the botanist wrote. "The noble timber of which there is such abundance would furnish plenty of materials for building either defenses, houses, or vessels; the river would furnish plenty of fish, and the soil make ample returns of any European vegetables, etc., sown in it."

And so it went. First came whalers and sailors, then traders and missionaries, then colonists. In 1840, Māori chiefs signed a treaty with the British Crown that made New Zealand a British colony. The treaty was supposed to settle the ownership of land, but then, later, when the Māori would not sell what the treaty said was theirs, there was war. The colonists took much of the land they wanted. When they tested whether their favored species would survive, they found that nearly all of them did—the climate of New Zealand was a close match to that of their native isles. "The English were frenetic introducers," Wink Sutton told me. The colonizing farmer-gardeners brought the usual wheat, corn, potatoes, cows, pigs, and sheep, along with more frivolous things, like red deer and rainbow trout and weasels.

In his journals, Joseph Banks several times noted the quality of New Zealand's native forests. "The woods . . . abound in excellent timber fit for any kind of building in size, grain, and apparent durability," he wrote. The colonists set upon the forests, felling them sometimes for timber but often for space. Where trees grew, they imagined sheep. They cleared the land even faster after the advent of refrigerated shipping in the 1880s, Sutton said. Now New Zealand could send its mutton, beef, cheese, and butter all the way back to the mother country. "It meant a huge expansion of land use," he said. "Forest was seen as an impediment."

The islands were thus sculpted to fit the colonists, emptied of bothersome trees and filled with sheep. But toward the end of the 1800s there arose a new worry: At the rate the islands' forests were disappearing, it seemed possible that they might soon disappear completely. At an 1896 New Zealand forestry conference, attendees agreed that they should begin searching for new timber trees. The islands' native trees grew too slowly to be relied upon, so the farmer-gardeners gathered up more than a hundred species from all over the world. Among the contestants in this timber tree olympiad were proven heavyweights like the Douglas-fir, European larch, ponderosa pine, and coast redwood. By all accounts, the Monterey pine was an afterthought.

In 1913, New Zealand's Royal Commission on Forestry issued a report on the nation's forests that summed up the result of its decade-old planting experiments. It is hard to know, the commissioners noted, what will be useful in the future—what types of timbers will people want, for instance, and for what purpose? But one thing was certain, they wrote: New Zealand would require good softwood timber for structural and agricultural uses. "Here *Pinus radiata* comes first," they wrote.

This was a surprise. While the Monterey pine had been planted widely in New Zealand as an ornamental and as a windbreak, people saw its timber mostly as firewood. Now the commissioners heaped praise on the tree. "It thrives in every variety of soil," they wrote, "from that of rich alluvial valleys to stony river-bed and sand-dune. It will also grow on the steepest of dry clay hillsides." Planted in thick stands with others of its kind, it grew perfectly straight, nothing like the hunched form it often took in the seaside groves of California and Mexico. Its wood was suitable for fruit cases and framing timbers alike. Best of all, the commissioners wrote, "Its rapidity of growth is truly remarkable." Everywhere they looked, they saw places that could be planted with the "all-important *Pinus radiata*."

So it went. We drove through endless dark green pines, Wink Sutton providing commentary. He explained the physiology of Monterey pine and the economics of raising it; he is an investor in several ongoing Monterey pine ventures. He noted a cluster of large pines whose owners—foolishly, he thought—had let them grow too big. "The idea that it's a commercial species, that it has to be managed, is beyond most people," he said. Sutton spoke about the tree with the interest and enthusiasm of a dairyman on the subject of cows. He recited an almanac's worth of Monterey pine arcana, pausing only when other drivers honked at him because he'd veered into their lane or slowed down unaccountably. When this happened, as it did half a dozen times during our drive, he would fall into a cloudy silence, from which I had to coax him by asking questions about Monterey pine.

Finally we arrived at our destination, Brooklands Forest Park, the small farm that contained Sutton's museum. We pulled in and were met by a man even older than Sutton who wore a cowboy hat and a flannel shirt and a cow-shaped belt buckle. He stood in the middle of the drive, looking puzzled. Sutton rolled down the window and told the man that he knew the owner.

"Don't get lost," the old man whispered.

"I won't get lost," Sutton said. "I know this place very well." We continued past the houses and parked next to an irrigation pond, rousting a gang of sheep. We walked to a gate. Sutton had hoped it would be dummy-locked, but instead it was really-locked. Sutton took the chest-high barrier slowly. It had been a while since he'd scaled any fences, he said. We walked up a gravel road, the fence on one side, Monterey pines thick on the other. The ground underneath the trees was carpeted in their needles. "It's farther than I remember," he said, his cane tapping on the stones. "I guess I'm getting slower."

At last we came to a signpost that said "1900." The Monterey pines behind it grew in an impassable thicket, a dog-hair mess of spindly trees. Many of them were dead. Their branches and trunks knocked together ominously in the wind. Sutton was delighted. "Oh, good," he said. "That shows exactly what I wanted to show." We'd entered the museum. This was the first exhibit.

The commissioners of the 1913 report were impressed by the Monterey pine's speed of growth and its easygoing acceptance of stony river bottoms and sand dunes, but the tree had shortcomings. From the lumberman's perspective, the ideal tree is rectangular, more easily milled into right-angled boards. Branches leave unsightly knots, so the fewer the better. Its wood is either light or heavy, stiff or flexible, depending on the desired use. It is strong, attractive, and rot resistant. The Monterey pine was badly wanting by all of these measures. It was cylindrical (although no more so than most trees), had lots of branches, was of medium weight and low strength, and rotted easily. But the commissioners thought these problems could be fixed. "It is quite probable," they wrote, "that by means of selection an improved variety of the species may be established."

Wink Sutton's museum aimed to show this process of domestication.

Rowland Burdon, Bill Libby, and Alan Brown wrote in *Domestication of Radiata Pine* that there are two parts to domestication. The first is husbandry, the process of learning how best to raise a species. Foresters' goal is to produce as much wood as quickly as possible. For them, the trade-off is between density—trees per acre—and the effects of competition among trees. Sutton's exhibits showed this. They represented the planting and thinning regimes of 1900, 1920, 1949, and 1968. As we walked through the museum, gradually the dog-hair tangle of 1900 opened until, by the 1968 exhibit, it felt like a park, with wide spaces between the trees, each of them too big for me to encircle with my arms. The trees in all of the exhibits were the same age, all planted in 2002, within a couple months of one another.

The second part of domestication is genetic improvement. As foresters honed their planting methods, they also worked to better their stock, breeding for the fastest-growing, straightest trees. Today, the Monterey pines in New Zealand's plantations grow far faster than their wild relatives. Foresters have made less progress on other fronts. The tree still produces timber that is relatively weak, knotted, and vulnerable to rot. It also remains stubbornly cylindrical. But when I asked Sutton if he would choose Monterey pine again were he to start the selection process over, he was emphatic. "Oh yes," he said. "It's a very plastic species." It bends to the will of the farmer-gardener like few others. "You stupid, stupid bastards," he recalled telling some Corsican pines that had failed to recognize the favor he'd done them by thinning.

Places with conditions suitable for Monterey pines are perhaps rare on the central California coast and on the islands west of Baja, but they are not rare in the world. Foresters across the Southern Hemisphere came to the same conclusion as the New Zealanders. There are now extensive Monterey pine plantations in Chile, South Africa, and Australia (as well as Spain, in the Northern Hemisphere). Libby, Rowland, and Brown wrote that while the exact bloodlines of these antipodal trees are lost to time, it is likely that the seeds David Douglas collected were among the first to make the trip south; if not these, they wrote, then seeds from trees grown with Douglas's seeds. The Monterey pine is now among the most widely planted trees in the world. It grows over thousands of square miles, these plantations covering an area some five hundred times greater than that of the wild groves. Along the way, the tree got a new name. The name "Monterey Bay" carries little weight outside the United States, so people in the Southern Hemisphere now mostly call the pine by its Latin name, *radiata*. In Wink Sutton's Kiwi accent, it came out as *ray-dee-otter*.

Sutton and I arrived at a wooden deck, across from a large sign that read, "New Zealand Forestry Group Ltd, PLANTATION MUSEUM, Conceived and Designed by Dr. Wink Sutton, Established September 2002." Sutton had wanted to build—or rather plant—a museum like this one for decades, he told me. Finally the owner of this property donated a corner of his land to the effort. He ran a bed-and-breakfast on his farm and thought that Sutton's museum might make a nice diversion for guests. Sutton had in mind a grander future. At one point, he showed me where the tour buses would turn around. I asked him about the last time he'd visited the museum. It was five years ago, he said, when he brought out another reporter, a correspondent from a Kiwi forestry trade publication.

The museum, to put it kindly, was unlike any other I'd been to. Caught by only the gentlest of daydreams, you could have walked all the way through its exhibits without even noticing. But I tried to see it by the light Sutton intended, in a triumphant way, as rare documentation of how determined farmer-gardeners shaped a bit of the world to fit.

We stood on the deck, quiet for a moment, looking out at the various sizes and arrangements of radiata pines, at this simulation of a farm that looked like a forest. "Well," he said finally, "that's my radiata museum." Then we turned and followed the track back down through the woods.

In his *Voyage of the Beagle*, Charles Darwin made what today seems like an obvious point: Rarity precedes extinction. For there to be a *last* something, there must first be a *second-to-last* something. "To admit that species generally become rare before they become extinct," Darwin wrote, ". . . appears to me much the same as to admit that sickness in the individual is the prelude to death."

By then, the story of the Mauritanian dodo, which had disappeared in the middle of the seventeenth century, had already become shorthand for extinction. Darwin noted the way that gauchos were able to lure the docile Falkland Islands fox with meat held in one hand, then stab it with a knife held in the other. He predicted that soon "this fox will be classed with the dodo, as an animal which has perished from the face of the earth." (He was right.) Clearly, humans could exterminate an entire species.

What had caused earlier extinctions was less certain. Georges Cuvier, the Catastrophist, had argued that the world's climate and landforms underwent sudden, apocalyptic upheavals, in which thriving species might suddenly cease to exist and new ones might appear. By this view, extinction was inherently unpredictable. But Darwin thought differently. As species appeared, he wrote, so did they disappear, sliding from rarity into oblivion. What happened to the dodos was not unusual. The proximate cause of extinction might vary—an asteroid, a plague of rats, gauchos—but the ultimate cause was always a bad case of rarity.

Asa Gray echoed Darwin in his 1872 speech on sequoias. "A species limited in individuals," he said, "holds its existence by a precarious tenure." The only sure cure was to become less rare. Gray didn't think such a turnaround was possible for the sequoias where they currently lived. But he thought the trees had gotten lucky. They'd been visited first by the hunter Gus Dowd, then soon after by botanists who carried their cones around the world. Thanks to these botanists, Gray said, "the species will probably be indefinitely preserved to science, and for ornamental and other uses, in its own and other lands."

Gray seems to have been ahead of his time. Most early conservationists were concerned with wild, romantic Nature, with ancient groves and scenic vistas. In those instances where their interest delved to the level of an individual species, what they tended to care about was where the species grew wild. This is the type of conservation that a modern scientist would call *in situ*, Latin for "on-site." Ecologists use the phrase to mean "in the wild." By the measure of the early conservationist, members of a species that humans had carried to somewhere else tended not to count.

The farmer-gardeners who carried species around the world, testing where each species would grow and what it might be good for, were equally unconcerned with conservation. When the botanist William Lobb raced to the Sierra Nevada in 1853 to collect sequoia cones, he was only carrying out his role as a plant hunter, an employee of nurserymen who aimed to corner a market hungry for exotic conifers. In his journals, David Douglas often described the plants he encountered as rare, but he seems to have spent little time pondering what effect sending their seeds to England might have on the future of the species. So too with the foresters who later filled the Southern Hemisphere with radiata pines. However much they admired the tree, they did not act out of concern for the native groves.

Occasionally in the course of their explorations, plant hunters did save something. In North America, the most famous example is the Franklin tree. The botanists John and William Bartram discovered it during a 1765 expedition near the Altamaha River in eastern Georgia. It was late autumn. The tree had no flowers, and they could not identify it. In 1776, William Bartram returned to the area and found the tree in full bloom. "These large white flowers stand single and sessile in the bosom of the leaves, and being near together towards the extremities of the twigs and usually many expanded at the same time, make a gay appearance," he wrote. In the years between the first sighting and the second, he wrote, he "never saw it grow in any other place, nor have I ever seen it growing wild, in all my travels, from Pennsylvania to Point Coupé, on the banks of the Mississippi, which must be allowed a very singular and unaccountable circumstance; at this place there are two or three acres of ground where it grows plentifully." Bartram assigned it to a new genus, calling it *Franklinia alatamaha.* He collected seeds, which he carried back to Philadelphia. The tree is now planted widely as an ornamental. It has not been seen in the wild since 1803.

Gradually the interests of the plant hunters and the conservationists converged. Conservationists came to recognize, as had Asa Gray, that sometimes the act of moving a species could deliver it from existential danger. Scientists call this type of conservation *ex situ:* "off-site." People argued as early as the 1920s that botanical gardens should focus on cultivating rare species, wrote biologist Vernon Heywood, but it wasn't until the 1970s and '80s that the gardens came to widely heed this call. Following the example that zoos had set decades earlier, they began to rebrand themselves as institutes of research and conservation. As they had done for centuries, these gardens sent botanists out into the wild to gather plants.

Rob Nicholson was one of these botanists. Not a wandering botanist, though. When I used the phrase in reference to Nicholson's Enlightenment-era predecessors, he stopped me. Sure, he'd wandered. To find rare and unusual plants, he'd traveled around the world—to Africa, Asia, South America. But the actual collection, he said, "it's a pretty precise endeavor." You study your quarry in advance. You go to herbaria to learn where other botanists found it. You know exactly where you're heading before you ever set foot in the field. It's all in the legwork.

Even to a botanist practiced in the arts of hunting and gathering, the Florida torreya was a tough bounty, he said. You could wander around for days and never find the tree. It was the summer of 1989, and Nicholson was working for the Arnold Arboretum, in Boston. The arboretum had partnered with the Center for Plant Conservation, a nonprofit that helped botanical gardens and arboretums to collect and maintain *ex situ* populations of hundreds of rare species of plants. Nicholson got the job of collecting the Florida torreya, one of the rarest. "Torreya was always kind of the poster child for rare plants," he said, "because it was so endangered, and it was crashing."

Hardy B. Croom discovered the Florida torreya, *Torreya taxifolia,* in 1833. Taxonomists named it after the botanist John Torrey. Asa Gray was Torrey's student and assistant at the time. He was present when Croom brought Torrey a sample of his new namesake. In the spring of 1875, Gray went to see the tree for himself. "The people of the district knew it by the name of 'Stinking Cedar' or 'Savine,'" he wrote, "the unsavory adjective referring to a peculiar unpleasant smell which the wounded bark exhales. . . . I may add that, in consequence of the stir we made about it, the people are learning to call it Torreya."

As described earlier, the trees had a small range, growing only in the steep ravines that cut through the bluffs overlooking the Apalachicola River. The range extended twenty-five or thirty miles along the river and inland five or ten. Gray repeated his earlier assessment of the sequoias, writing that "any species of very restricted range may be said to hold its existence by a precarious tenure." But the torreya was locally common. Gray found an abundance of the trees, of all ages and sizes. Unless the area was "wantonly disforested," Gray wrote—and he thought the steepness of the terrain would prevent that—"this species may be expected to endure."

It did, for a while. But then sometime in the 1950s, the tree began to decline. The June 8, 1962, issue of *Science* magazine included a letter to the editor from two professors in the biology department at Florida State University, Herman Kurz and R. K. Godfrey. The letter was titled "The Florida Torreya Destined for Extinction." Kurz had visited Torreya State Park in 1954. At that time, he and Godfrey wrote, "No one present noticed anything abnormal about the trees." Now, though, "there remain but a scattering of skeleton trunks, a few of which have abortive sprouts at their bases." The cause of the torreya's decline seemed to be a fungus, although it was unclear what caused its sudden attack on the trees. Maybe the recent stretch of dry years had stressed the trees, enabling an endemic fungus to grow out of control. Or maybe the fungus was exotic and had been introduced when the land surrounding the park was turned over to longleaf pine plantations. Whatever the cause, Kurz and Godfrey wrote, "It is unlikely . . . that any corrective measures can be taken to preserve the Florida torreya in its native soil."

In 1983, the U.S. Fish and Wildlife Service proposed that the Florida torreya be listed under the Endangered Species Act. The stated purpose of the act, which was passed by Congress in 1973, is to "provide a means whereby the ecosystems upon which endangered species and threatened species depend may be conserved." One of its key powers is to designate and protect "critical habitat." In the case of the Florida torreya, though, Fish and Wildlife recommended against trying to designate critical habitat for the tree. "The wild trees do not now have good long-term survival prospects," it wrote. The best shot for Florida torreya was *ex situ*.

At that time, planted torreyas were few and scattered. "If you look at the history of botanic gardens, it was almost like a stamp-collecting mentality," Rob Nicholson said. "One of each is good enough." The farmer-gardener approach of the past didn't reflect what scientists now know: that a species is the sum of its genes. The ideal *ex situ* population contains every variation of a species' genes. As a collector, that means taking samples from trees scattered across the range of the species.

That was the hard part. Nicholson told me he would have struggled to find any of the trees, were it not for Mark Schwartz, a postdoc at Florida State University, Gainesville. "He'd spent weeks and months and years exploring around in these ravines," Nicholson said. "He knew where they all were." Nicholson and Schwartz spent a week hiking the bluffs. "It's like a treasure hunt for grown-ups," Nicholson said. "You're looking for that one organism in this big, crazy mix of plant species, then you finally find it, and it's like, 'Ah, we did it.'" They took thousands of cuttings from 163 torreyas. At the end of the week, Nicholson flew back to Boston.

Nicholson recalled that on this flight, he was "for some weird reason" bumped up to first class. "'Wow, this is so great, what a treat,'" he remembered thinking. "I'm sitting next to some, like, business executive–type guy, and we get talking: 'What do you do?' 'What do you do?'" Nicholson, a man with a flair for the dramatic, said he told the businessman that, down in the cargo hold, he had a cooler holding "thousands of cuttings of this plant that's probably going to go extinct." The businessman pondered this. "You could see him thinking cost-benefit analysis," Nicholson recalled. "And he says, 'Well, do we have to save every species?'"

Nicholson was ready. In his younger days, he explained, he'd spent a year working in an airplane engine factory. He was a riveter. "Look out at the wing of the airplane," he told the businessman. "You see all those rivets? Imagine they're species, and they're falling out one at a time. Tell me when you want to stop." I'd heard this one before. Biologists Paul and Anne Ehrlich told it in their 1981 book, *Extinction*. Nothing is tempting like a good parable. The businessman was impressed. "That kind of drove it home to him," Nicholson said. "The future is unwritten. You try to save as much of the natural world as you can. You don't know which one will be crucial."

"I lived it," he added. "I used to rivet."

Back at the Arnold Arboretum, Nicholson worked to prepare the torreya cuttings. He sprayed them with fungicide and doused them with rooting hormones. "They root easily, thank God," he said. When the cuttings were rooted, he shipped them to ten arboretums and botanical gardens across eastern North America and Europe. Over the following years, many of these cuttings died, victims of their minders' inattention or inexperience, Nicholson said. "It takes a particular set of skills to grow a plant in a captive environment." One person was especially successful in raising the trees: Ron Determann, a horticulturist at the Atlanta Botanical Garden. "Ron Determann will go down as the guy who saved torreya," Nicholson said.

In the winter of 2018, I visited the Atlanta Botanical Garden's auxiliary site in Gainesville, Georgia, with Emily Coffey, a paleoecologist and the garden's vice president of conservation and research. In one corner of the fenced lot, a thicket of torreyas grew under black shade cloth stretched over wooden lath. The netting was to protect the trees from the Georgia sun. Florida torreya is an understory species. Nicholson, in an article published in *Natural History* magazine in 1990, wrote that the wild Florida torreyas were likely doomed. But he envisioned a new refugium, "an artificial *Torreya* forest where pollen can float, genes mingle, and the evolution of the past hundred million years can continue, even if it is in a pitifully discounted format." The *ex situ* grove was just what he imagined, Nicholson told me. "Ron [Determann]'s plantation, producing more genetically diverse seed than could be found in the wild, is exactly what I had hoped for."

The point of an *ex situ* collection is not to look nice. Still, it can feel like an odd kind of salvation. The torreyas of Gainesville were pitiful indeed. They were crowded together, bent and twisted under the shade canopy, sick from the same fungus that afflicted the wild trees. I was reminded of the dog-hair pines in Wink Sutton's radiata museum.

In a 1998 paper on the Monterey pine, Forest Service paleoecologist Connie Millar proposed a third option, a point midway between the *in situ* and *ex situ* poles. The five native Monterey pine populations of California and Mexico, she wrote, were beset by urbanization, drought, disease, the effects of fire suppression, and—on the island of Guadalupe—a horde of feral goats. In addition, people had carried *Pinus radiata* seeds back from New Zealand and planted them alongside the native groves, threatening to muddy the genetic differences among the five native populations and reduce the pine's overall genetic diversity. Had the tree really gotten lucky? In turning the tree to their purposes, farmer-gardeners now threatened to permanently change it. The future of Monterey pine in its native groves was more in doubt than ever.

But this depended on the definition of "native," Millar wrote. At the same time that people scattered Monterey pine across the Southern Hemisphere, they had planted the tree in other parts of California. "In many coastal areas it has survived and thrived," she wrote. Monterey pine is inarguably a novel addition to New Zealand. It is not native there. Things were less clear in California, though, Millar wrote. The fossil record showed that during the Pleistocene, Monterey pine had roamed up and down the coast. Many of the places in California where farmer-gardeners had planted Monterey pines were within that historic range. With this in mind, people could take a broader view of nativeness, she wrote. "Intentional and accidental planting may be viewed as an assistance to Monterey pine in achieving conditions that natural migration might have promoted." She proposed calling these populations "neo-native."

As of the early 2000s, no similar third option existed for the Florida torreya. Aside from the native groves along the Apalachicola River and the *ex situ* collection at the Atlanta Botanical Garden, there was no place where torreyas existed together in any numbers, no place they had been planted where they could be plausibly regarded as wild.

But then, in the winter of 2004, a science writer named Connie Barlow visited Torreya State Park. She lay under one of the suckering torreyas, looked up through its boughs, and told it she would move it.

Scientists realized early that living things in general—and plants in particular—might not keep pace with modern climate change. "Even if suitable land is preserved for a species to shift into," Robert Peters and Joan Darling wrote in 1985, "extinction may still occur if present habitat becomes unsuitable faster than new habitat can be colonized." The migration routes that plants had followed in the past might now be cut off, wrote paleoecologist Margaret Davis in 1989. Rare species would be in especially great danger. "Preserve managers will have to learn to propagate plants," she wrote, "transplanting them to areas that are predicted to become favorable for them and attempting to preserve genetic diversity in the face of diminishing natural populations." In 1992, biologists Richard Primack and S. L. Miao wrote that "plant conservation biologists should at least consider trying to assist rare species to shift their ranges slightly northward, to somewhat higher elevations or to more suitable local environments, in response to global climate change."

When Connie Barlow first began thinking about moving the Florida torreya north, she was unaware of this burgeoning scientific discussion. The idea occurred to her while she was working on a book about "ecological anachronisms," plants that coevolved with animals that had since gone extinct. In *The Ghosts of Evolution*, she examined dozens of these pairings: the avocado, missing its giant, armadillo-like glyptodonts; the Osage orange, missing its mastodons; Mauritania's tambalacoque tree, missing its dodos. In these partnerships, the fruit is a fleshy lure, the animal a wandering seed dispenser. Without their animals, the plants were marooned.

Barlow thought the Florida torreya had all the characteristics of a plant that had gotten stuck. Since the mid-1800s, naturalists and horticulturists had theorized that the tree was out of place. In 1872, Asa Gray wrote that the tree "seems as if it had somehow been crowded down out of the Alleghanies into its present limited southern quarters." In 1905, John Coulter and W. J. G. Land wrote in the *Botanical Gazette* that "the conclusion is irresistible that Torreya is a northern plant." In its 1986 recovery plan, Fish and Wildlife wrote, "The species may be restricted to the area because it failed to migrate northward at the end of the Pleistocene." It was what scientists call a "relict," an organism living in a small fragment of a once-greater range. To Barlow, the situation seemed obvious: The ravines along the Apalachicola River were not the only places on Earth where the Florida torreya could survive but, rather, the only places where it could survive that it had also managed to reach. Recent warming had favored the fungus and disfavored the tree. The problem wasn't so much disease, she thought, as climate. "It wants to head north," she wrote, "but it hasn't found a vehicle."

In 2002, she read a new book by paleobotanist Hazel Delcourt, called *Forests in Peril*. Delcourt echoed scientists' earlier predictions: Many modern ecosystems would collapse in the face of climate change. The quilt of species would be disassembled into its component patches and resewn. People might not like what this new quilt looked like. The new quilt might be smaller and less comfortable than the old one. Humans would have to work hard to have any chance of preserving familiar ecosystems. "If we can predict where the loss of biodiversity will be greatest, we can provide corridors to allow for species to migrate," she wrote. "We may also need to be prepared to transplant endangered species to new locations where climate will be favorable." Barlow told me that *Forests in Peril* was the final bit of encouragement she needed.

She wrote to paleoecologist Paul Martin. He was the author of the Pleistocene overkill theory and, in a 1982 paper coauthored with ecologist Dan Janzen, had coauthored the idea of ecological anachronisms. By the early 2000s, Martin was pushing a new idea. In *The Ghosts of Evolution,* Barlow quoted him telling a South Dakota audience, "I want to leap ahead of the understanding that we finally accept extinction as a phenomenon, to the possibility of doing something about extinction." The missing megafauna could be replaced, he said, not exactly, but in approximation. Where mammoths and mastodons had once roamed North America, imported Asian and African elephants would suffice. Wild horses would suit in lieu of extinct horses, and extant camels could approximate extinct ones. An African lion would be almost as good as the vanished American lion. This "rewilding," Martin said, could restore entire anachronistic landscapes.

Now, Barlow told him, they could do the same thing with the Florida torreya. Martin liked the idea. Barlow compiled a list of possible co-conspirators. It included conservationists, scientists, and managers of several botanical gardens. Hazel Delcourt was on the list, as were Rob Nicholson and Mark Schwartz, who had started the *ex situ* population of Florida torreyas. Stan Simpkins, a U.S. Fish and Wildlife employee who oversaw federal conservation efforts of the Florida torreya, was included as well. Barlow laid out her plan in an e-mail to the group. Few of the participants seemed comfortable with the idea of moving the tree north.

"It is not clear to me that there is strong evidence for the idea that moving the species north will automatically result in healthy populations," wrote one.

"I am skeptical that this represents a viable conservation strategy," wrote another.

One person who followed the discussion with interest was Joshua Brown, an editor of the now-shuttered environmental magazine *Wild Earth*. "I have followed the exchanges about the potential translocation of *T. taxifolia* with great interest," Brown wrote. Barlow proposed that the magazine print a pair of articles, "maybe like Supreme Court decisions, with a Yes and a No, and individuals then weighing in with their own nuances to each."

Wild Earth published the resulting articles in its fall–winter 2004–2005 issue. Barlow and Paul Martin wrote the *Yes* opinion. Planting the Florida torreya in the southern Appalachians would be merely returning it to ground it likely held before the Pleistocene, they wrote. It was in one sense exactly the same as what people had been doing for thousands of years, the same as what David Douglas did when he sent Monterey pine seeds to England, and yet, at the same time, it was something quite different. Barlow and Martin called it "assisted migration," a term coined by graduate student Brian Keel in 2002. "With Florida torreya," they wrote, "we can explore the ecological and social dimensions of what seems likely to be a radically new era for plant conservation."

The debate that eventually followed was not over the underlying facts: Many species would be unable to keep up with the changing climate; some of those species would face extinction. The debate was over the solution. In a 1996 paper, Mark Schwartz anticipated this debate's important points. Many rare plant species of the American Southeast—including the torreya—would likely be left behind by the shifting climate, he wrote. This did not pose "an unsurmountable conservation problem." People could help the plants keep up with the changing climate. But Schwartz urged caution. "Conserving historically accurate representatives of natural communities is, perhaps, the most common management directive and a common conservation goal," he wrote. "Intentional introduction of species perceived to be vulnerable to extinction into new habitats compromises traditional conservation efforts." Moving one species could upset the arrangement of many others.

KISS YOUR ASH GOOD-BYE

Once there was a Frenchman named Étienne Léopold Trouvelot who lived in Medford, Massachusetts. Trouvelot was a lithographer and astronomer by trade, but his hobby was insects. He wanted to start an American silk industry. "Every one is familiar with the beautiful and delicate fabric made from the fibres spun by that crawling repulsive creature, the silk worm," Trouvelot wrote in an 1867 article in the *American Naturalist.* North America, he noted, was home to many "gigantic species of moths" whose caterpillars might produce fantastic amounts of silk. But most of them had disqualifying flaws. The silk of caterpillars in genus *Callosamia* was "so strongly gummed" that Trouvelot could not unspool the cocoons. The silk of genus *Platysamia* could be easily unspooled but lacked brilliance, and "the worm is too delicate to be raised in large numbers." The silk of *Tropea luna*, "the magnificent green moth," was thin and brittle. Finally Trouvelot found success with the polyphemus moth, *Antheraea polyphemus*, an enormous brown moth with sleepy blue eyespots. He reported that he'd spent the last six years learning to raise the caterpillars, which are the size, shape, and color of an electrified pickle. He described his efforts in polyphemus husbandry, concluding, "*—To be continued.*"

But Trouvelot did not continue his account. Soon after the first installment was published he suddenly lost his taste for moth breeding. Sometime in the late 1860s, for reasons that are unclear, he'd imported the eggs of the European gypsy moth. Some reports say that he wanted to breed the moths with an American species to create a hybrid. Maybe he was just interested in them. Hobbies sometimes stray beyond reason. But it is rare that a hobby goes so badly and publicly awry.

The gypsy moths escaped. How this happened is also unclear. Some sources claim that the caterpillars escaped out a window of Trouvelot's house, while others say a storm freed the caterpillars from an outdoor tent. However it occurred, Trouvelot knew that it might be a problem. The authors of an 1896 report wrote that his neighbors recalled that he was "much disturbed on being unable to find them."

Merde.

The farmer-gardeners carry species around the world, offering new ways out of dead ends and bridges over once-impassable barriers. Some of these species we mean to carry: wheat and sheep and Monterey pines. Many others are bycatch: We do not want them, really, but haul them aboard anyway. Purposeful or not, each moved species is an experiment.

It is easy to say, in the broadest sense, why a species is in a place: It can survive the physical conditions in that place and can coexist with the other living occupants of that place, plus it managed to get there. But it's harder to say why a species is not in a place. Is it because physical conditions aren't suitable? Is it because it couldn't compete there, or because other species killed it off? Or is it only because it hasn't managed to arrive? This ambiguity means that we often cannot predict which things will succeed in which places. It means that when we move living things from one place to another, sometimes they will succeed beyond our wishes.

Trouvelot thought that the gypsy moth might succeed. It was a pest in its native Europe, where it ate hundreds of species of trees and shrubs. At first it seemed his fears were unfounded. For a decade after the gypsy moths escaped, all was quiet in Trouvelot's neighborhood. But then, toward the end of the 1870s, the caterpillars reappeared. They marched down country roads like an invading army, stripping bare any plants they could find.

The sound of their chewing kept people awake at night. Townsfolk raked the caterpillars into piles and burned them, but the pyres did no good.

"The caterpillars would get into the house in spite of every precaution, and we would find them even upon the clothing hanging in the closets," wrote one resident.

"A mere shake of a tree would bring them down on one in showers," wrote another.

"The stench in this place was very bad," wrote a third.

Despite efforts to contain the outbreak, European gypsy moths spread out from Medford, Massachusetts. They are now established across the northeastern United States, as far south as Virginia, and west to Wisconsin. They continue to spread. Occasionally their numbers explode, and they eat holes in the forest. "It's wicked bad here," the mayor of a small town in eastern Connecticut told the *Hartford Courant* in the summer of 2017. "Wicked bad!"

Many thousands of species now grow in places they hadn't reached on their own. Mountains are flattened, deserts skirted, oceans hopped. Through this grand reshuffling, we have proved beyond doubt that where a species lives in the wild is not always where it is able to grow biggest and fastest or reach greatest abundance. We have proved that sometimes all it takes to save a species from extinction is to move it.

But Connie Barlow and Paul Martin's proposal to move the Florida torreya north struck many people as naïve. Barlow and Martin presumed to know not only why the tree was languishing in its native groves, but also what it would do when they sent it somewhere new. Species are unpredictable. Each new species in a new place is an experiment. Sometimes experiments go badly.

In recent years, people have called the world of reshuffled species the "New Pangaea," in reference to Pangaea (that's "all Earth" in Greek), an ancient landmass that contained all the world's modern continents. All of the world's land-dwelling creatures lived there together. Some 180 million years ago, Pangaea split into the Laurasia and Gondwana supercontinents, then into the smaller pieces that became the world's modern continents. In life, separation is what drives creation. Individuals of a species are isolated from others of their kind, either in space or in habit, and eventually one species becomes two. Farmer-gardeners now reunite the long-separated threads of evolution. In this story of humanity's heavy footprint, the Anthropocene is the time, and New Pangaea the place. But if the message of the Anthropocene is simply that people have touched every hollow on Earth, every crevice, then the message of New Pangaea is that we can still make things worse.

By a modern reading, Léopold Trouvelot's choice of hobby seems eccentric, but it would not have seemed unusual at the time. In the mid-1800s, "intercontinental transplantation of species was all the rage," wrote historian Peter Coates in *American Perceptions of Immigrant and Invasive Species*. Along with ranks of enthusiastic hobbyists like Trouvelot, many countries had acclimatization societies, whose members hoped to improve their surroundings by the addition of new species from around the world. The mission of the Cincinnati Acclimatization Society, Coates wrote, "was to 'introduce to this country all useful, insect eating European birds, as well as the best singers.'" Even governments participated in the reshuffling. As plant ecologist Mark Davis noted in *Invasion Biology*, the original purpose of the U.S. Department of Agriculture, created in 1862 by President Abraham Lincoln, was to "procure, propagate and distribute . . . new and valuable seeds and plants."

People have long known that these species, once transplanted, sometimes got away. Swedish naturalist Pehr Kalm traveled through the American colonies in the 1740s. He encountered many species of European origin, among them clothing moths, bedbugs, mice, rats, and cockroaches. (In his journal, he recalled a Swedish immigrant named Sven Laock, who "told me, that he had in his younger years been once very much frightened on account of a cock-roach, which crept into his ear whilst he was asleep.") Charles Darwin noted in *On the Origin of Species* that these new arrivals sometimes pushed the old inhabitants out. "Several of the plants now most numerous over the wide plains of La Plata, clothing square leagues of surface almost to the exclusion of all other plants, have been introduced from Europe," he wrote. Frogs introduced to the islands of Madeira, the Azores, and Mauritius, meanwhile, had "multiplied so as to become a nuisance."

These transplanted species were more than mere nuisance, wrote George Perkins Marsh in his 1864 book *Man and Nature*. "Every plant, every animal, is a geographical agency," he wrote. "Whenever man has transported a plant from its native habitat to a new soil, he has introduced a new geographical force to act upon it, and this generally at the expense of some indigenous growth which the foreign vegetable has supplanted." Marsh was an American polymath—a lawyer, legislator, diplomat, and speaker of some twenty languages. In *Man and Nature*, he argued that humanity had made important changes to Earth's geography, often for the worse. Marsh described the global destruction of forests, the draining of wetlands, the digging of canals and raising of dikes, along with the erosion, flooding, desertification, and extinction that often followed. Humans had rearranged the Earth, Marsh wrote, and so had they rearranged its species. It was another type of destruction. A few years after Marsh published *Man and Nature*, Léopold Trouvelot's gypsy moths escaped into the wilds of Medford, Massachusetts.

For species like the Monterey pine, the farmer-gardeners meant a gift of whole new continents. For other species, though, the farmer-gardeners created the new possibility of things going suddenly and catastrophically wrong. In these stories, the focus is not on the new arrival, but on the new arrival's victims. In the summer of 1904, Hermann Merkel, the chief forester at what is now the Bronx Zoo, noticed dying leaves on one of the zoo's chestnut trees. The massive American chestnut, *Castanea dentata*, was the "wild forest king," wrote Susan Freinkel in *American Chestnut*.

In *1491*, Charles Mann wrote that "in colonial times, as many as one out of every four trees in between southeastern Canada and Georgia was a chestnut." The chestnut's timbers built Appalachia. Its seeds fed the great clouds of passenger pigeons and both the Native Americans and the European colonists who displaced them. The American chestnuts in Merkel's care were dying of chestnut blight, *Cryphonectria parasitica*, a fungus native to closely related Asian chestnut species. People fought to stop the fungus from spreading, but by the mid-1900s, it had swept across the trees' entire range, killing nearly all of them.

The story repeated in 1931, when Curtis May, a U.S. Department of Agriculture employee, found Dutch elm disease on several American elms in Cleveland and Cincinnati. The disease had been discovered a decade earlier in Europe, after millions of elms died in the final days of World War I. People first suspected the die-off to be an effect of shelling or nerve gas, but eventually traced it to a fungus spread by beetles. The elm had been a symbol of the American project, alongside the Stars and Bars and the bald eagle, wrote Thomas Campanella in *Republic of Shade*. People planted millions of the trees along streets across the country, where their arching boughs formed "a verdant parasol soaring above the quotidian, casting it in a dappled and flattering light." By the 1970s, almost all of the big elms were dead.

American attitudes toward foreign species soured, wrote Peter Coates. "Enthusiasm waned toward [the nineteenth] century's end as the unanticipated drawbacks of certain promising introductions . . . became increasingly apparent." Assimilation societies fell into disrepute, and Americans took to calling even such once-loved species as the English sparrow "lazy little louts," "vagrants," and "wretched foreigner[s]"—mirroring in language, at least, their prejudices against foreign-born humans. By the early twentieth century, the USDA had begun working to keep native species in and nonnative species out.

But new species kept arriving. International trade ballooned in the decades after World War II, the farmer-gardeners' long quest to find the world's most useful, exotic, entertaining, and cheapest things ripening into full-on globalization. This trade meant more chances for species to arrive, and more chances for them to escape.

In 1958, British ecologist Charles Elton published *The Ecology of Invasions by Animals and Plants.* "It is not just nuclear bombs and wars that threaten us, though these rank very high on the list at the moment," he wrote. "There are other sorts of explosions, and this book is about ecological explosions." In the book, Elton noted dozens of these "ecological explosions," among them Étienne Trouvelot's gypsy moths. "A hundred years of faster and bigger transport has kept up and intensified this bombardment of every country by foreign species, brought accidentally or on purpose, by vessel and by air, and also overland from places that used to be isolated," he wrote. "We must make no mistake: we are seeing one of the great historical convulsions in the world's fauna and flora." Once a curiosity, the escapees had become a central trouble.

By now, these stories of New Pangaea are so familiar as to be predictable: A new creature arrives, and though it is docile in its homeland, in its new surroundings it swells into a plague, covering the land, fouling waterways, gobbling down native species, annihilating a favored tree. Try as we might, we cannot stop its spread. We are surrounded by versions of this story. But it can be hard to know why species live where they do and not where they do not. It can be hard to know what shackles we undo in moving them. Which species will do well in which places can never be fully known, so the same story still manages to surprise, over and over.

This is perhaps the best way to explain why David Roberts did not immediately know what was happening on the day in June 2001 when an arborist asked him to take a look at some ash trees on the grounds of a condominium complex in Plymouth, Michigan, just west of Detroit. Roberts is a plant pathologist at Michigan State University, in East Lansing.

On his website, he calls himself "the Tree Doctor." He gets these kinds of calls often, someone wanting his opinion on some sick plant.

At the condos, Roberts found dozens of large ash trees, all with dead leaves and witches'-brooms—clusters of sprouts bursting from around the base of their trunks. This didn't seem unusual. Southeastern Michigan had just endured several years of drought. There were many sick trees. Roberts thought at first that the ash trees had suffered a chemical burn—maybe the gardeners had applied fertilizer too close to their roots. But the gardeners told him they'd done no such thing. Roberts moved down his list of suspects. The next likely culprit, he figured, was ash yellows, a bacterial disease that had afflicted ash trees across the Northeast for decades and had recently reached the Midwest.

Ash yellows often causes witches'-brooming, but the trees Roberts visited were missing the stunted, yellowed leaves that give the disease its name. When I talked to him, he pointed out this and other ill-fitting pieces of the puzzle. But back then, he said, there was no reason to suspect the worst. "It's easy, looking back now, to say, 'Why didn't we see that before? Why didn't we do something about it?'" he told me. "I thought it was just going to be another one of those things that's a dead end." Doctors don't always know what's wrong with their patients. Sometimes things get better on their own. Roberts took a few photos of the trees and left.

At the time, North America's ash trees seemed safe, secure, of least concern. The United States alone is home to sixteen species of ashes, genus *Fraxinus*. Ash was one of the most common trees across the Midwest. By one estimate, Ohio alone held nearly four billion ash trees. Ash was a kind of background tree, neither so insistently majestic as a giant sequoia nor so notably unkempt as a black spruce. Its bark is checkered or furrowed, and it has compound leaves arranged like dog teams, with one lead-dog leaflet out front and four or six or eight leaflets trailing behind in pairs down the petiole. Each leaflet is a saw-toothed spearhead. In form, most species of ash are stereotypical—the kind of cloud-on-a-popsicle-stick tree a kid might draw.

It is an especially useful tree. Its wood is strong and tough and flexible, the stuff of snowshoe frames, sled runners, and tool handles. The haft of Achilles's spear in *The Iliad* was ash, wrote Robert Penn in *The Man Who Made Things Out of Trees*, a history of ash. So were Babe Ruth's Louisville Sluggers. It is the wood of canoe paddles and railroad ties, of flooring and electric guitars, of cabinetry and lobster traps. Native Americans across the Midwest and East craft intricate baskets from black ash, *Fraxinus nigra*. Ash tolerates the trials of the floodplain, where it often lives in the wild, and also the urban abuses of flood and drought, of pollution, of road salt, of cramped roots. This agreeable character made it a popular street tree across much of the United States. Some three hundred million ash trees grew along the streets of Michigan. What had especially endeared ash to city forestry planners was that it had been almost completely free of pests. Nothing would kill it.

A month after his first trip to the condominiums, Roberts made a follow-up visit. He found a crew cutting down the sickened ashes. On one of the trees the bark had fallen away from the wood, and he saw for the first time what would soon be a familiar sight: a mess of squiggling tunnels through the phloem. Later, he showed me pictures. Tunnels crossed and recrossed, covering the surface of the log like a plate of spaghetti. "I'd never seen that much tunneling," Roberts said. Some of the tunnels still had larvae in them—each an inch long, off-white, with bell-shaped segments. The tunnels encircled the tree, cutting off the flow of water and nutrients, choking it. The witches'-broom sprouts from the base of the trees were lifeboats, the roots' desperate last effort to hold on after losing the main trunk. Roberts also found tiny holes in the trees' bark, each shaped like a capital D.

Roberts showed the grubs to his entomologist colleagues at Michigan State. He said they told him the larvae looked like that of a drab black beetle called the two-lined chestnut borer, *Agrilus bilineatus*, a native pest of American chestnut trees. It is a member of a large genus of beetles whose larvae burrow and feed on the sapwood of trees and other flowering plants. The beetles' larvae pupate under the bark, then emerge as adults, leaving the characteristic D-shaped exit hole Roberts noticed on the ash trees, matching the beetles' flat bellies and rounded backs.

Roberts told me that this identification didn't quite make sense. *Agrilus* beetles are usually picky eaters, confining their diets to plants that are closely related. After the destruction of American chestnut trees, the chestnut borers had moved to oaks, which are in the Fagaceae family, the same as chestnuts. But ash is an entirely different family—for the chestnut borer to attack it would be an unusual culinary leap. (Roberts's entomologist colleagues, for their part, told me that they would not have attempted, on the basis of larvae alone, to identify the *Agrilus* beetle to the level of species; this is one of several points where Roberts's telling of events differs from that of his colleagues.) The damage the beetles were causing was strange for another reason: *Agrilus* beetles tend to be opportunistic, attacking trees that are already weakened by drought or other damage. They aren't usually killers on their own.

But Roberts didn't make too much of these problems back then. He still thought the real culprit was ash yellows. During his trips around the southeastern corner of the state, he noticed more ash trees with similar symptoms. He learned to spot the ailing trees from a distance, picking them out by the light-colored patches on their trunks where woodpeckers had chipped away bark to dig out grubs. In the spring of 2002, he collected leaves from these sick trees to test for ash yellows. At the same time, he harvested logs from a few of the trees that had already died. He brought the logs back to his lab in East Lansing and wrapped them in plastic garbage bags. He hoped to rear out the adult beetles, which, he assumed, would be two-lined chestnut borers.

A few weeks later, Roberts got the test results back. The problem wasn't ash yellows. Now the beetles were impossible to ignore. One day he went back to his lab and found them flying around the room. They'd emerged from the logs and chewed their way through the plastic garbage bags. It was then that all of the hints and mislaid pieces fell together. He realized that when he'd visited the dead ash trees at the apartment complex in Plymouth, he'd encountered something new. "I'm not by any means a good entomologist," Roberts said. "But I had not seen this borer before." The beetles were members of the genus *Agrilus*, no doubt—blunt-headed and the size of a grain of rice—but instead of the chestnut borer's drab black, the beetles Roberts found in his lab were a brilliant emerald green.

On a summer morning in 2017, I met David Roberts at a Starbucks in Canton, Michigan. He has an ambling, folksy manner, and long hair and a beard. He wore mirrored aviators and a blue dress shirt unbuttoned to reveal his luxurious chest hair. He looked like a latter-day Bee Gee, an impression little dimmed by his green MSU Spartans baseball cap. As word of the strange beetle spread, Roberts began leading tours of the area for entomologists and members of the state and federal agriculture departments. The tours started there at Starbucks. As we walked to my car, Roberts pointed out a row of Bradford pear and honey locust trees at the edge of the parking lot. They were replacements, he said. "These used to be all ash trees over here."

Roberts took the green beetle he found in his lab to his entomologist colleagues at MSU. They didn't know what it was. Nor did the entomologists at Oregon State, where they sent the beetles next. Nor did their colleagues at UCLA. No one in the entire United States had seen the beetle. Finally, someone sent a specimen to Eduard Jendek, an entomologist in Bratislava, Slovakia, and the world's leading authority on *Agrilus* beetle taxonomy. By July 2002, Jendek had an ID. The beetle was *Agrilus planipennis*, first described in 1888. It was from eastern Asia. The only information on the species was a brief write-up in a Chinese forestry textbook from 1992. The text gave the barest life history of the beetle and named its host tree as *Fraxinus chinensis*, the Chinese ash.

It felt like the opening of a familiar story. "We're searching the literature, and all we're finding are taxonomic reports, really nothing about lifestyle," said Deb McCullough, an entomologist at Michigan State. "It's invasive, but nobody knows how to survey for it. We don't know anything about its biology. The shit hit the fan."

Roberts and I drove toward Plymouth. When Europeans first arrived in Michigan, nearly the entire state was forested. Settlers cleared forests for farms, and timber companies and a series of large fires took much of the rest. A state report from the late 1920s estimated that by the time intensive timber harvests stopped, less than 10 percent of the state's forests remained. In the decades since, the portion of the state that is forested has crept back up to roughly half. The southeastern corner, which includes Ann Arbor and Detroit, is the least-treed part of the state, but even there, the forests are the defining—really, the only—feature. The trees themselves were indiscernible to me as we sped by—my eyes are of the American Northwest, channels tuned conifer. I couldn't pick anything up in the endless hardwoods. The underlying land was flat, too, planed smooth by Pleistocene ice. The built environment was no help, all chains and strip malls. Driving through look-alike towns, I felt directionless, the visual equivalent of having cotton in my ears. Rumpled clouds provided the only relief.

Roberts directed us to the Bradbury Park Homes condominiums. The condos were single-story, red brick and off-white clapboard, with trimmed lawns and tidy gardens. The streets between had a yellowish hue that lent the scene a sepia tinge. "This is it right here," Roberts said, pointing across the street at one of the houses, where an American flag was flying. "See that flag?" There had been an ash growing in the yard, he said, the first one he inspected when he visited back in 2001. We knocked on the door, but the woman who answered said she was a new arrival. She didn't know about the ash trees. Her neighbor couldn't tell us much, either. "Nobody knows anything about anything anyway," she said as she closed the door. Finally we found a woman named Barb, out planting begonias and marigolds in her garden, who remembered when the grounds crew cut down all the ashes. Barb was impassive. "We had to," she said. "They were infestering, or whatever."

The entomologists I spoke with remember that time more vividly: as a mad scramble, trying to understand and outmaneuver the beetle. Deb McCullough told me that during one of the tours with Roberts, they decided to give the new beetle a common name, something the newspapers could print. They ruled out "green ash borer," because green ash was the common name for a specific tree, *Fraxinus pennsylvanica*, and the ash borer seemed to attack all sixteen species of North American ashes. "Asian ash borer," meanwhile, would invite xenophobia. "Big ash borer" was no good, either—"People would slur their words," McCullough explained. The name the Michiganders settled on came from a joking reference to *The Wizard of Oz*, she said. "We were driving through some not-great parts of Detroit, and we said, 'We're on the yellow brick road, searching for the emerald ash borer!' "

Roberts's tours quickly became unnecessary. It turned out that the emerald ash borer had already reached nearly all the ashes in the southeastern corner of the state. In the summer of 2002, the Michigan Department of Agriculture enacted a quarantine, banning the movement of ash logs, nursery stock, and firewood into or out of five counties surrounding Detroit. But immediately there were questions about whether the boundaries of the quarantine were in the proper place. Was the borer truly contained? The people trying to stop the beetle needed a way to detect it.

At the MSU campus in East Lansing, I visited the lab of Therese Poland, an entomologist with the U.S. Forest Service. She took me into a basement room with mustard-yellow tiled walls, linoleum floors, and black countertops piled with machines. She pointed out one of these, a box with a number panel on one side. It was an expensive type of oven. She used it to create an ash borer lure.

Leaves emit many different scents. The question was which scents were characteristic of ash trees, and which would attract ash borers. Poland first collected the odor of ash leaves by pumping air through bags of leaves and into tubes of Super Q powder, an absorbent. She then rinsed the powder with pentane, a solvent. She collected the scent-infused liquid in glass vials. Then she used an instrument to inject tiny amounts of this liquid into the oven. As she slowly raised the heat, volatile compounds from the leaves vaporized at different temperatures. These compounds flowed from the oven to a Y-shaped splitter, with half going into a mass spectrometer, which recorded the ions that made up each compound.

On the computer screen, these compounds appeared as spikes, like the wiggles on an earthquake seismogram. The other side of the tube led to the "electro-antennal detector," which, Poland explained, was where the decapitated antennae went. A freshly severed beetle antenna will keep responding to its favorite scents for hours, she said, "just like how a chicken runs around with its head cut off for a while." When a volatile compound matched one of the antennae's receptors, the antennae sent an electrical impulse toward the body—or, in this case, where the body would have been. A computer recorded both the molecules and the electrical impulses from the beetle's feelers. When a new molecule appeared at the same time as the antennae fired, she explained, "Then you say, 'Oh, that's the compound I need!' "

At the other end of the room, she dug a plastic bag out of a cluttered freezer. The bag was filled with a chemical called cis-3-hexanol, also known as green leaf alcohol. She cracked the bag. Out wafted hints of gasoline and cut grass, potpourri and pepper. It smelled terrible to my nose, but the beetles like it. Researchers have spent the last decade tweaking the recipe, and the lures now reliably attract ash borers. But there are fewer ash trees now. Back in the early days of the fight, when the forest was still thick with ashes, the lures couldn't compete. The scientists would catch ash borers, but not until after the surrounding woods were completely infestered.

Other efforts to stop the invasion were similarly troubled. The Michigan Department of Agriculture enacted its quarantine in 2002. In theory, the quarantine should have stopped all movement of ash nursery stock, timber, and firewood in the affected counties. But in practice the quarantine proved difficult to enforce, especially the ban on moving firewood. "A lot of it was through education and outreach," said John Bedford. "We couldn't search every car." Bedford worked for the ag department, directing more than one hundred surveyors. They spent the summer of 2003 working their way across southeastern Michigan, checking ash trees for signs of the borer—cracked bark, dying leaves, witches'-brooms, and the D-shaped exit holes. By the end of 2003, they'd found the beetle in thirteen counties in Michigan, and in Ohio. By 2004, the Department of Agriculture was cutting down healthy ashes. It hoped to create impassable buffer zones around the edges of the invasion. Canadian foresters soon followed suit, felling every ash tree in a six-mile-wide swath that extended from Lake St. Clair, near Detroit, down to Lake Erie, a fuel break they thought would stop the beetle's spread. No sooner would foresters complete these barriers, though, than they'd find the borer on the other side.

If anything, the beetle seemed to spread more quickly as time went on. The ash borer was discovered in Illinois and Maryland in 2006, Pennsylvania and West Virginia in 2007, and Wisconsin, Missouri, and Virginia in 2008. The entomologists and officials knew they were losing the fight. The story was going the usual way. "It was rough," McCullough said.

"I used to be excited for it to be winter, because then you couldn't tell if the trees were alive or dead," Bedford said. "It was extremely depressing."

In 2014, a group of scientists led by U.S. Forest Service researcher Nathan Siegert published a study on the origins of the emerald ash borer invasion. They had stitched together a chronology of the insect's spread using cores they collected from thousands of dead and dying ash around southeastern Michigan. They found that the epicenter of the invasion was in the town of Canton, not far from the condominiums I visited with David Roberts.

Deb McCullough, a coauthor of the study, told me that given the ash borer's life cycle and habits, she thought the beetle had likely arrived hidden in a wooden pallet, the kind used to stack goods for international trade. These pallets are often made of ash. The area around Canton is home to multiple automobile parts factories. In the wake of the disaster some people blamed these factories.

After our visit to the condominiums, Roberts and I drove by a factory owned by a Japanese auto parts manufacturer. We rolled slowly past the big red warehouse, snapping photographs, then turned into its shipping and receiving lane. "Somebody said there was a huge pile of pallets out back," Roberts said, but when we got there we didn't find any—just a few trailers and some dumpsters. On the way back out to the main road, though, I noticed a trailer with "WE BUY RECONDITIONED PALLETS" written on its side in big letters. I slowed the car to a crawl as Roberts leaned out the window with his camera. It was kind of fun, but whether this factory was individually responsible for the invasion probably misses the point. Siegert and his coauthors found that the beetles had likely arrived in Michigan by the early 1990s, maybe as long as a decade before an arborist called Roberts to ask him about the dying trees.

In a 2000 paper, Forest Service entomologist Sandy Liebhold wrote (adapting a concept from a 1986 paper by A. P. Dobson and R. M. May) that invasions have three stages. The first is arrival. How the emerald ash borer arrived is unsettled. Maybe it came in on a pallet, as McCullough suggested, or maybe on a live, nursery-bound Manchurian or Chinese ash, as entomologist Jian Duan told me he thought was more likely. However it got here, the beetle's population was small at first. It was rare, still vulnerable. The second stage of invasion is establishment, which "can be considered the opposite of extinction and represents the growth of a newly arrived population sufficient such that extinction is impossible," Liebhold wrote. The third stage is spread, "the process by which the species expands its range into the new habitat." By the time Roberts found it, the emerald ash borer was established and spreading. It was already too late.

The time to stop an invasion is before the species is established, or better yet, before the species even arrives. This is the mission of various branches of the U.S. Department of Homeland Security and the Department of Agriculture. Their agents inspect fruits, vegetables, live plants, seeds, passengers, animals, luggage, ships, planes, shipping containers, pallets, dunnage, and other sundries, ensuring that all are in compliance with a web of national laws and international treaties. Various items and organisms are fumigated, heated, cooled, or quarantined.

These agents did not find the emerald ash borer. Inspections reduce the risk of invasion, Sandy Liebhold told me, but can't eliminate it. The problem is volume. In the century and a half since Étienne Léopold Trouvelot imported his gypsy moths, international trade has exploded. The volumes in question are voluminous indeed: The amount of cargo carried by ships increased by an average of more than 3 percent each year between 1970 and 2010, reaching 10.7 billion tons in 2017, with planes carrying another 66 million tons. That year, planes also carried more than 4 billion passengers. People import as many as 3 billion live plants into the United States each year. Each ship, each ton of cargo, each passenger, each plant means a potential ride for an invader.

Invasion biologists sometimes refer to a rough rule of tens: For every ten species, just one will manage to arrive somewhere new. For every ten species that manages to arrive in a new place, just one will survive and establish. For every ten that become established, most will fade into the background, doing no noticeable damage. Just one will go on to be a damaging invader. Biologist Dan Simberloff told me that the rule of tens focuses on economic damage rather than ecological damage, and so likely underestimates the size of the problem. But the broader message holds: The odds of a future invader arriving on any given ride are slim, but these slim odds are multiplied by the vast number of potential rides.

Hiring more inspectors won't solve the problem, Liebhold said. "There's a misperception that inspection by itself is a primary mechanism by which we exclude potentially dangerous organisms," he told me.

"It's not. We could have ten times as many inspectors and we're still going to miss most of the dangerous pests." The agents I spoke with at APHIS (that's the U.S. Department of Agriculture's Animal and Plant Health Inspection Service) agreed. "You can't find everything," said Ron Komsa, one of the agents. The purpose of inspection is to get a sense of which organisms are arriving which ways, so that inspectors can focus more attention on those routes. The more times a given organism arrives, the better the odds that it will become established. Biosecurity merely aims to lower the odds.

Even focused this way, the system is by its nature one of reaction. Inspectors are constantly trying to keep up. Take fruit flies and blueberries, Komsa told me. It used to be that blueberries grew in cold places, like Maine and Quebec, and fruit flies grew in the tropics, and never the twain did meet. But then blueberry growers bred blueberries to grow in Brazil, and all of a sudden American biosecurity inspectors turned up blueberries infested with fruit flies. It's the kind of thing that happens constantly. The problem is always evolving.

Far better would be a system of prediction. The basic problem of biological invasions is that each new invader comes as a surprise. Which of any given ten species will arrive in a new place? Which of any ten arrivals will survive? Which of any ten survivors will spread? Which might cause harm to things that humans care about? "We're not very good at predicting which ones will cause problems and which ones won't," Dan Simberloff told me. "The Holy Grail of invasion biology is predicting impact." For more than a century, naturalists and scientists have puzzled over this: What is it, exactly, that makes an invader?

A few patterns present themselves. The most obvious is that a species that is invasive in one place will often be invasive in another. But there are many exceptions, Simberloff said. Similarly, the close relatives of weeds are likely to become weeds themselves. But again, there are exceptions; as Asa Gray wrote in 1876, "In many cases it is easy to explain why a plant, once introduced, should take a strong and persistent hold and spread rapidly. In others we discern nothing in the plant itself which should give it advantage." Glenda Wardle, an ecologist at the University of Sydney, Australia, pointed out to me that in its native land, the Monterey pine gave few hints of its future success. Now many of the countries with extensive radiata pine plantations are also fighting to control "wildling pines," escaped from the farm and spreading out into the country.

Another obvious pattern is that organisms that are able to move quickly over wide distances are more likely to become invaders than slower, less mobile organisms; so too with organisms that reproduce quickly. But these traits are so common as to be nearly useless as a way of winnowing potential invaders. Islands are more likely to be invaded than continents. A species that has traveled far is more likely to become an invader than one that has traveled only a short distance. A species with a way of life unlike that of any species in its new surroundings is more likely to be an invader than one whose way of life is similar to that of its new neighbors. But Simberloff said that in all cases, there are long lists of examples where the pattern doesn't hold.

There is one especially striking pattern in the history of forest invasions. Over and over again, North America's trees have been attacked by insects or fungi that are native to those trees' sister species. North America's ashes succumb to a beetle native to Asian ashes, American elms to a disease native to Asian elms, American chestnuts to a disease of Asian chestnuts. These pairings are a legacy of the ancient temperate forest that lived in the space now inhabited by the boreal. As the world grew colder in the Oligocene and Miocene epochs, this forest went south, splitting among North America, Europe, and Asia. Each species was isolated from others of its kind. Now each continent has its own oaks, maples, elms, chestnuts, pines, firs, yews, spruces, hemlocks, beeches, poplars, rowans, and many other congeners—members of the same genus. But this pattern is still too broad to be useful in specific cases. Among them, these trees are hosts to thousands of fungi and insects. Most of these organisms, given the chance, do not prove deadly to their host's congeners. "These patterns only get us so far," Simberloff said. Put a species in a new place, surrounded by species that are new to it, and you never know quite what will happen.

Then there is the more earthbound problem of—to borrow the former defense secretary's durable phrase—the unknown unknowns.

From Michigan, I traveled on to Washington, D.C., where I visited the Smithsonian Museum of Natural History's entomology department. My guide was a man named Gary Hevel, curator emeritus, bearded, with beetled brow and wide black suspenders. Hevel pulled out several boxes of his favorite, flashiest insects: creatures with dinosaur horns, pneumatic pincers, fly-rod antennae; with wings patterned like Persian rugs, like stained glass, like the flush of the setting Sun; beetles with polka dots and stripes and chevrons, with shells of corrugated tin, of hammered bronze, of gasoline on water. He showed me a moth whose back was emblazoned with the visage of a human skull, and a butterfly that he told me lepidopterists hold to be the most beautiful of insects, its wings painted with an archipelago of deepest black in a metallic ocean that shifted from green, turquoise, and sage to yellow, orange, red, purple, and blue. In one of the boxes was an emerald ash borer, a polished green bullet the shape and size of a grain of rice. I stared at it. If you didn't know what it was, your eye would skip right over it.

Hevel led me through the collection. The galleries between the metal cabinets smelled overwhelmingly of mothballs, although Hevel claimed that the smell was "much better" than it used to be. We reached the *Agrilus* section, and Hevel pulled out drawer after drawer, all filled with the ash borer's sister species. Most of the beetles were roughly the same size and shape as the ash borer. Most were variations on a dull brown theme. Genus *Agrilus* took up three cabinets, each of which held a dozen drawers that in turn held hundreds of specimens representing dozens of species. We walked through rooms full of insects awaiting identification: bail-top jars of pickled grubs, flies in alcohol, beakers of assorted invertebrates. Labels were of varying specificity: "Madagascar, *Coleoptera*," "Spangler, Mexico, 1964," "Miscellaneous Water Beetles." Who knows what lies in those catacombs. Field collectors often fumigate whole trees, then gather up the insects that drop onto tarps spread out below. Maybe someday a graduate student will come pick through them.

"No one is likely to get into New Zealand again accompanied by a live red deer," Charles Elton wrote in *Ecology of Invasions*. Intercontinental transplantation of species is no longer "all the rage," as Peter Coates put it. Releasing strange animals and plants into the wild is frowned upon. But the potential for accidents is still high. Entomologists estimate that there are more than five million species of insects, about a million of them described by science, and only a handful studied in any detail. There are untold millions more fungi, bacteria, and viruses. Scientists are turning up new species constantly. Most of the entomologists I've spoken with have discovered new species; Gary Hevel alone has discovered nearly four hundred new species.

After the tour, Hevel took me down the hall to meet Mike Gates, a USDA taxonomist who specializes in Hymenoptera, the order of insects that includes wasps and ants. Gates showed me several packages he'd just received in the mail. He ripped one open and pulled out a vial of alcohol containing one small ant. He handed me the accompanying form. Inspectors had intercepted the ant at the airport in Charlotte, North Carolina, on a flight from Mexico. It was in someone's baggage, in a container of "unknown leaves." Gates receives packages like this every day from inspectors all over the country. Each package is a question: *Do you know what this is? Should we be concerned?*

After a century of trying to quell the invasive tide, the suggestion that we ought to be rewilding or assisting migrations or otherwise moving species into new places struck many scientists as ill-considered. "Humanity has a long record of tinkering with natural ecosystems," wrote Mark Schwartz in 2004. "Largely these have been successful from the perspective of the human endeavor—think agriculture." But these successes often had dire consequences for the world's other living things. There is danger in moving species around, he wrote. "It is not an action to be taken lightly."

Schwartz is the ecologist who helped Rob Nicholson begin a Florida torreya *ex situ* collection in 1989. He was writing in *Wild Earth*, in an article published alongside Connie Barlow and Paul Martin's proposal to move the Florida torreya north. Schwartz's article was the rebuttal. Scientists had been suggesting the idea of assisted migration for nearly two decades, but these suggestions had garnered little attention and prompted little debate. Schwartz's article, titled "Conservationists Should Not Move *Torreya taxifolia*," was the first to make an explicit case against the idea.

Schwartz agreed with Barlow and Martin that the tree was a glacial relict, marooned after failing to migrate north on its own. The tree, he wrote, is "quite likely on the edge of its climatic tolerance, and might do well in a cooler climate." Modern climatic warming would probably make the situation worse. It might be okay to move Florida torreya, he wrote. While there are many stories of trees becoming invaders—tamarisk, paperbark, and tree of heaven are a few of the most notorious in North America—Schwartz thought that "the likelihood of *Torreya taxifolia* expanding out of control is low." The tree was slow-growing, with big, heavy seeds that "are a favorite food of squirrels," he wrote. "[It] carries all of the attributes of a species that will not spread and become a noxious weed." However deserving the Florida torreya seemed, though, Schwartz was opposed to moving it.

For one, he wrote, there was the question of why the tree was dying. The fungus that seemed to be killing it had appeared in the 1950s. Barlow thought that the fungus was not an invader but a native, and that its sudden virulence was a symptom of the tree's climatic mismatch. But Schwartz was skeptical that merely moving the tree north would relieve these symptoms. At the time, nobody knew what the fungus was or where it had come from. "It is not clear what Florida torreya would be escaping from," he wrote.

In 2011, a group of researchers led by Jason Smith, a plant pathologist at the University of Florida, Gainesville, reported that they'd identified the fungus that had caused the Florida torreya's long decline: a previously undescribed canker fungus in the *Fusarium* genus. In a follow-up paper, five researchers led by geneticist Takayuki Aoki named the fungus *Fusarium torreyae*. The researchers suggested that the fungus was of east Asian origin, although this has not yet been proved.

Jason Smith and other scientists I've talked with think this fungus is the Florida torreya's central problem—and that it has the potential to cause a similar problem for other trees. In 2012, Smith's graduate student, Aaron Trulock, reported the results of his experiments infecting other types of trees with *Fusarium torreyae*. He tested the Florida torreya's sister species and other species of conifers. Some of these species currently live alongside the torreya in Florida. Other species were native farther north, in the Appalachians. Several trees in this latter group appeared susceptible to the fungus: Fraser fir, red spruce, eastern hemlock, white pine, and Table Mountain pine. Moving the torreya north, Trulock concluded, would threaten other native species. "You are risking a whole other ecosystem there," Jason Smith told me. "I think that's irresponsible."

Barlow dismissed these concerns. Trulock's research was conducted in laboratory conditions that may or may not match conditions in the wild, she told me. His research also lacks credentials as a scientific document; it is only a graduate thesis, not a peer-reviewed paper published in a journal. Furthermore, by the time she began her effort to save the Florida torreya, the people running the official conservation effort had already sent seeds and cuttings around the world, to botanical gardens as far away as Scotland. She said they were just as likely to have spread the fungus. "Was it a gamble we could spread disease?" Rob Nicholson said. "Possibly, but we felt there was no other good alternative."

Mark Schwartz's main concern back in 2004 was not the fungus, though. It was that assisting the northward migration of the Florida torreya would set a dangerous precedent. The tree might be benign, unlikely to turn invader. Moving it north might even save it. But what about the next species someone thought needed help? "Societal recognition of an appropriate reticence toward species introductions has been slow, but is emerging," Schwartz wrote. The invasion biologists' admonishments were slowly sinking in. This idea of assisted migration threatened to undo that progress. Someone might once again think of carrying red deer to New Zealand.

But Schwartz admitted that there was little anyone could do to stop Barlow and Martin from moving the trees north. The Florida torreya was listed under the Endangered Species Act, which made it illegal for anyone to collect or trade in seeds or cuttings from the wild plants. But the act contained a loophole: Seeds or cuttings from cultivated plants could be collected and traded. Several nurseries offered Florida torreyas for sale. "Anyone who wants to plant Florida torreya," Schwartz wrote, "can do so—wherever they want."

That, more or less, is what happened. At the end of their 2004 *Wild Earth* article, Barlow and Martin wrote that people interested in volunteering should visit the website of the group they'd formed, the Torreya Guardians. That fall, Bill Alexander gave the Torreya Guardians some 110 seeds. Alexander was the forest historian at the Biltmore Estate in Asheville, North Carolina, which had a small grove of Florida torreyas. Guardians member Lee Barnes, a horticulturist, sent the seeds to a dozen volunteers scattered across the eastern United States and to several botanical gardens in Europe. The next year, the Biltmore gave the Guardians 200 seeds, and 300 the next.

Fossil evidence of genus *Torreya* in North America is sparse. As Paul Martin and Connie Barlow wrote in their proposal to move the Florida torreya north, "The most recent macrofossils identified as the genus *Torreya* in eastern North America are upper Cretaceous"—that is, between 66 and 100 million years old. Some of the places the Torreya Guardians sent the tree likely hadn't seen the Florida torreya—or any other torreya—for tens of thousands of years, or even millions of years. In some places, it was likely entirely new.

From Washington, D.C., I continued on to New York City and Connecticut, and finally to Millbrook, in New York's Hudson Valley, where I met an ecologist named Gary Lovett at the Cary Institute of Ecosystem Studies. On the deck outside the institute, Lovett pointed out what looked like bread crumbs scattered around us. "These are gypsy moth poops," he said. "But I think they must have just swept the deck." There was less poop than usual. We walked toward a nearby hemlock tree, a conifer with short, soft needles and graceful drooping boughs, in this case festooned with what appeared to be white plastic grocery bags. The bags were actually a mesh fabric, Lovett explained as we got closer. They contained flies. Scientists from the University of Vermont were testing whether the flies would eat hemlock woolly adelgids, little sap-sucking bugs that feed on the hemlocks' needles, eventually killing them. The adelgids arrived almost a century ago, probably carried over on an Asian hemlock bound for someone's garden. The flies were worth a shot, Lovett said, although the odds of success were low; people had been trying and failing to stamp out the adelgids for many decades. It will probably go the same way with the emerald ash borer, he said. "I think we can probably kiss our ash good-bye, as they say."

By now, emerald ash borer is one of the most destructive invaders ever to hit North American forests. At this writing, the beetle has reached thirty-six American states. It is also in five Canadian provinces and is spreading across western Russia, where it attacks planted North American ashes and the European ash, *Fraxinus excelsior*. In 2017, the International Union for Conservation of Nature, which tracks species around the world that are at risk of extinction, added six species of North American ashes to its Red List. It described the Carolina ash as endangered and five others as critically endangered: white ash, green ash, blue ash, black ash, and pumpkin ash. In 2002, when David Roberts found the emerald ash borer flying around his lab, all six had been common.

On the day of my visit, Gary Lovett was preparing for a trip to Washington, D.C., where he was going to try to convince members of Congress to enact the recommendations of a program called Tree-SMART Trade. In 2005 and 2006, the United States adopted an international set of standards that imposed stricter controls on wooden pallets, dunnage, and other wooden packing material, requiring that it be either heat-treated or fumigated before crossing international borders. But the measure had proved only partly successful, Lovett said. His group wanted to discontinue the use of all wooden packing materials in international trade. The group also wanted to restrict imports of all woody plants that share a genus with a North American native. Lovett said he would personally like to go even further, banning all imports of live plants. "Why are we importing live plants at all?" he said. "We have plenty of plants in North America." Lovett admitted these were lofty goals, sure to meet opposition from industry. Indeed, Patrick Atagi, a spokesman for the National Wooden Pallet and Container Association, later told me that the proposed restrictions were unnecessary, not to mention impractical. "Pallets move the world," he said.

In a paper on the history of plant quarantines, USDA risk analyst Christina Devorshak quoted G. A. Weber, who in 1930 wrote of a new biosecurity law that "few if any acts of the Department of Agriculture have aroused so much discussion and so much adverse criticism and condemnation on the one hand and commendation on the other." The tension "between protecting a country from the spread of pests, while not implementing undue restrictions on trade remains with us today," Devorshak wrote. Invaders are biological pollution, Lovett said, a cost of globalization paid by the public. But few legislators seemed to share Lovett's concern. "Given the current political climate, I don't know how far we're going to get," he told me, adding, "This shouldn't be a partisan issue, because trees aren't really partisan creatures."

We continued past the fly-festooned hemlocks, down a path through the woods, and arrived at a bench that overlooked a brook. Light dipped through drooping boughs. I could have pictured Emerson strolling by, or a pair of children leaving a trail of bread crumbs. But the only bread crumbs were really caterpillar poop, and the only ghosts were of vanished forests. Colonists cut these woods over at least once, Lovett said, and the chestnuts that grew back in were gone, taken by blight. The hemlocks that had replaced them were likely doomed, too.

It is strange and new for a forest to change this way, over and over—strange and new and yet also already mundane. I wondered aloud how many times this story of New Pangaea might be repeated, feeling a little sad about it. "If you asked me what the forest around here will look like in fifty years, I couldn't tell you," Lovett said. "Have you heard the one about the airplane rivets?"

I had.

Our efforts to slow the flood of invaders have had some effect. While the volume of trade has gone up exponentially over the last hundred years, the rate of new introductions of nonnative insects has remained relatively steady. "If we had no prevention whatsoever, you'd expect them on parallel tracks," he said. "It means we're stopping more than we used to. It's not as bad as it could be."

The tap might be shut, but the faucet still drips. New insects trickle into the United States, a couple each year. Every decade or so, one arrives that causes real damage. In 2011, a group of researchers led by Juliann Aukema estimated that, based on the present rate of species arriving, there was a 30 percent chance of an insect as damaging as or worse than the emerald ash borer arriving within a decade. By the time you read this, perhaps it's already here.

5

COUNTERPEST

By the time of my visit to Michigan, one part of the story of the emerald ash borer was long over. The beetle had hopped every barricade, dodged every quarantine. Ash—the sturdy, dependable, ubiquitous background tree—was gone. Even the voids it left seemed to be fading from memory. But another part of the story was still underway—indeed, still in its early days.

On a rainy, windy morning, the day before my trip to the Plymouth condominiums with David Roberts, I met Leah Bauer, Toby Petrice, and Therese Poland in a park about an hour north of Ann Arbor. They are entomologists with the U.S. Forest Service, based out of Michigan State University, in East Lansing. They work across campus from David Roberts, the plant pathologist who found the emerald ash borer. (Poland was the one who showed me the beetle-antennae apparatus described in the previous chapter.) All three were wearing hard hats. "The government is really hot on safety," said Bauer. By her bearing and her tendency to spin verbal donuts over the others' neat conversational lanes, she seemed to be the group's leader. One time when she was out in the field, a tree really had fallen on her, she said. "It snapped and fell straight in my eye." No lasting damage, luckily. They had forgotten to bring an extra helmet for me.

Bauer pulled a lunch box from the back of her car and unzipped it. The lunch box was filled with clear plastic cups. She held one up. "Anyway, so we just happen to be releasing this new little species," she said. Inside the canister were a dozen fuzzy specks, like little bits of luft that had not yet coalesced into full-on dust bunnies. The specks were wasps. Bauer had told me about them on the phone, but somehow they were even smaller than I'd imagined. I pointed out that these wasps really were tiny.

"Oh, these are huge," Bauer said, then laughed.

"Wait until you see the egg parasitoids," Petrice said. "You'll be like, 'Wait a second, where're they at?'"

The reason we were there, about to release a lunch box full of wasps into the woods, was because the wasps have a bloodhound's skill for finding emerald ash borer larvae buried under the bark of ash trees. The entomologists told me they weren't actually sure how the wasps do it—maybe they detect the vibrations the larvae make as they wriggle through the cambium, or maybe they pick up whatever scent the larvae give off. The wasps (on close examination) had long antennae and Victorian waists and tiny needle-like appendages extending from the end of their abdomens, thinner than an eyelash. This needle, Petrice told me, was a guide that contained a still smaller needle called the ovipositor. In other wasps, this organ evolved into the more familiar stinger, but these wasps use it to poke through the bark of the tree and lay their eggs in the beetle larvae.

"It's like a saw thing," Bauer said, when I asked how the wispy organ could possibly poke through tree bark. Petrice, meanwhile, compared it to an impact drill. It didn't make much sense either way. The entomologists assured me the important point was that the wasps get the job done. We doused ourselves with mosquito repellant and headed into the woods.

To say that the emerald ash borer is destroying North America's ash trees only begins to capture the extent of the damage. In the years after the emerald ash borer swept through midwestern forests, scientists tried to tally all that had been rearranged. Gaps appeared in the forest's canopy, and the ground was littered with ash logs. Oaks and maples and basswoods grew faster, benefiting from the extra light and space. The gaps the ashes left filled in with invasive shrubs. Swampy places grew swampier, without ash trees to transpire their water skyward. The types of bacteria that lived in the soil changed. The increase in sunlight hitting the forest floor made the plants that giant swallowtail butterfly caterpillars feed on harder to digest. Invasive earthworms increased in abundance.

Worst hit were the arthropods, members of the taxonomic order that includes insects and spiders. Entomologists Kamal Gandhi and Dan Herms tallied some 282 species of arthropods that either fed or bred on North American ash trees. Forty-three were known to feed or breed only on ash trees. These species faced extinction, Gandhi and Herms wrote. Their extinctions could cause other extinctions. "We . . . anticipate cascading effects on other affiliated species, including fungal, bacterial, and invertebrate associates and parasites," they wrote.

Humans, too, depended on ash trees. Ashes supported entire industries and were among the most widely planted trees in cities across the United States. One group of researchers estimated that removing and replacing dead ash trees could cost as much as $26 billion in Michigan, Indiana, Illinois, and Wisconsin. Another group of researchers estimated that between 1990 and 2007, the death of ash trees in fifteen states "was associated with an additional 6,113 deaths related to illness of the lower respiratory system, and 15,080 cardiovascular-related deaths." A third group of researchers found that the removal of ash trees in Cincinnati, Ohio, was correlated with an increase in crime.

The beetle is also destroying an important cultural resource. For millennia, Native Americans across the East and Midwest have woven baskets out of strips of black ash, *Fraxinus nigra*. Kelly Church, a member of the Gun Lake Band of Pottawatomi and Ottawa tribes, comes from a long line of basket weavers. Her people made baby baskets, fishing baskets, firewood baskets, berry baskets, and later, market baskets, she said. "There's always been a use for baskets." Church told me that from her home near Hopkins, Michigan, she used to be able to "go fifteen minutes in any direction, go get a tree, and be back in a half hour. Now we have to drive to the U.P."—Upper Peninsula—six or seven hours each way. But the beetle will likely soon claim the trees there, too. Church and her fellow weavers in Michigan have been gathering black ash seeds in the hope that they might someday replant.

Moving one species risks the arrangement of many others. Species live where they do and not where they don't because of the suitability of the physical settings, because of their interactions with other living things, and because of their ability to reach otherwise suitable places. What constrains a species in one place is often inscrutable, and how it will behave in a new place, unpredictable. "The equation of animal and vegetable life is too complicated a problem for human intelligence to solve," George Perkins Marsh warned in *Man and Nature*, "and we can never know how wide a circle of disturbance we produce in the harmonies of nature when we throw the smallest pebble into the ocean of organic life."

But whether the risk is worth it depends on the species involved. Assisted migration is not the only conservation method that involves moving species to new places and releasing them into the wild. For decades, scientists have worked to repair the damage that invasive insects and diseases have done to North America's forests. Many of these efforts involve carrying species to new places and releasing them into the wild. People greatly valued ash trees and chestnuts and elms, so they tinker where they otherwise might not and move things they otherwise wouldn't.

The Forest Service entomologists wandered through the woods, looking for good places to release the wasps. What I said before was true: Ash as it existed before the beetle *was* gone. But the woods were still filled with ash trees. They were shrunken, damaged, sorry things—what the entomologists called the "orphan cohort," the trees that grew from the last flush of seeds dropped by the old ash trees before they died. The biggest of those orphan ashes, with trunks perhaps four inches wide, bore squiggling scars from past attacks, their bark knotted up like varicose veins. "This would be a good tree here," Toby Petrice said, pointing to one. Its bark had light patches from where woodpeckers had chipped away layers of bark, and around its base was a witches'-broom of small sprouts, a sign that it hadn't yet succumbed.

Leah Bauer pulled a cup from the lunch box and loosened its lid. "You'll see the insects fly very quickly when you release them," she said. She pulled away the lid. I saw nothing, but when she held up the jar, half of the wasps were gone. A few that had died in transit remained in the jar, along with some living stragglers that seemed uneager to leave. She tipped the dead wasps out into another container so she could track how many she'd actually released. "And these guys that are still alive," Bauer said, "we'll just kind of put them on the tree and hope that they say, 'Hey, there's an ash tree.'" She held the cup over an ash leaf and banged the bottom of it. It gave a hollow *bop*. "To get them out of the cup, you have to give them a smart tap," she said. *Bop bop bop*. When she held up the jar again, all the wasps had vanished. There in the woods of Michigan, one insect from eastern Asia was reunited with another.

Everything has enemies. If there are too many rats, get a cat. If there are too many cats, get a dog. If there are too many dogs, get some tapeworms. It's an old idea. In A.D. 304, the botanist Ji Han wrote that the inhabitants of southern China "sell ants stored in bags of rush mats. . . . If the [citrus] trees do not have this kind of ant, the fruits will all be damaged by many harmful insects and not a single fruit will be perfect." Invasive pests could be controlled the same way. In *Man and Nature*, George Perkins Marsh wrote that the only way "to balance the disproportionate development of noxious foreign species" was to bring "from their native country the tribes which prey upon them." This practice of using one creature to check another is now called biological control. The practice of reuniting two associated species in a new part of the world, meanwhile, is called classical biological control. The goal is to place the invader back in the fetters that once held it.

Any species might swell into a plague, given the chance. "Linnaeus," Darwin wrote, "has calculated that if an annual plant produced only two seeds—and there is no plant nearly so unproductive as this—and their seedlings next year produced two, and so on, then in twenty years there would be a million plants." By this same math, a pair of elephants—"the slowest breeder of all known animals"—would become fifteen million in a mere five centuries. This geometrically increasing rate of population growth, Darwin wrote, "must be checked by destruction at some period of life." The world is not overrun with elephants because not every elephant that is born survives long enough to produce more elephants.

Living things are constrained by physical conditions and by the presence or absence of other living things. The presence of the emerald ash borer, for example, seems to severely curtail the space where North American ash trees are able to survive. The question is why Asia's native ash trees are able to persist even in the insect's presence. Perhaps it is because some other living thing keeps the emerald ash borer's numbers in check. Perhaps, as the USDA scientists hope, it is a tiny wasp.

In *Silent Spring*, Rachel Carson lauded biological control of insects as an alternative to pesticides. Chemicals are clumsy, tending to splash death beyond the edges of their target. A biocontrol agent (as modern scientists call these creatures), on the other hand, has tools perfectly built for the task of attacking the invader. "The advantages of [biological] control over chemicals are obvious," Carson wrote. "It is relatively inexpensive, it is permanent, it leaves no poisonous residues."

Charles Elton called these biological control agents "counterpests"—more gracefully, I think. "This highly technical field of activity has reached very wide proportions," he wrote in *The Ecology of Invasions*, "and there is now continual traffic of introduced counterpests to every country of the world that has any crop-growing or forestry." For example, he wrote, between 1953 and 1955 authorities imported sixteen counterpests into Hawaii: beetles from Mexico and Central America; moths from California and Brazil; wasps from Guam; and two carnivorous snails, one from

Florida and one from the Mariana Islands. Their targets were plants, flies, beetles, and another snail, this one vegetarian. "It is quite an exchange and bazaar for species," Elton wrote, concluding the list, "a scrambling together of forms from the continents and islands of the world."

Leah Bauer has spent much of her career studying counterpests. She's worked on gypsy moth counterpests and Asian longhorn beetle counterpests and spruce budworm counterpests. After the discovery of the emerald ash borer in Michigan, she began looking for creatures native to North America that might attack it. She and her colleagues found about a dozen insects and pathogens that attacked the beetle, including six species of parasitoid wasps, which laid their eggs on or inside the ash borer larvae or eggs. (A parasite subsists on another creature but doesn't usually kill it, while a parasitoid usually does.) Bauer and her colleagues found the parasitoid wasps by cutting down infested ash trees and stripping them of bark, then raising the ash borer larvae and eggs they found. Some of the larvae turned into adult ash borers, but others died and sprouted wasps. These wasps were of the right genre, but they were less aggressive than desired. While they killed a few ash borers, it wasn't enough to check the beetle's runaway population. Woodpeckers were more effective. "They're the most dominant natural enemy," Bauer said. "But we can't mass-rear them."

In 2003, Bauer's colleague Houping Liu traveled with Toby Petrice back to China to look for the creatures that attack emerald ash borers in their native lands. During the three-week trip, Liu, Petrice, and Chinese entomologists dissected about a hundred ash trees, either removing patches of bark with a drawknife or cutting an ash down and skinning the whole tree. Twelve of the trees they dissected were infested with emerald ash borers. From these twelve trees they collected 521 ash borer larvae, which they shipped back to Michigan State University, where Therese Poland and Leah Bauer had a quarantine lab. "We got kind of lucky," Liu told me—of the 521 ash borer larvae, 22 contained parasitoid wasp larvae. The wasps were of two species, both new to science.

A few days after my walk in the woods with Leah Bauer, Therese Poland, and Toby Petrice, I visited the U.S. Department of Agriculture's parasitoid-wasp-production facility in Brighton, Michigan. I found it in a drab industrial park, down the block from Terminix Termite & Pest Control. A pair of USDA Animal and Plant Health Inspection Service (APHIS) employees, Sharon Lucik and Ben Slager, led me through the facility. The first room was filled with shelves, which held cardboard barrels the size and shape of beer kegs. Every year, USDA employees cut down hundreds of donated ash trees that are infested with emerald ash borers. In return for their trees, landowners receive an equivalent quantity of seasoned firewood. The infested logs stay in cold storage to keep the emerald ash borer larvae in hibernation. Employees pull them out as needed and stick them in the tubes. The tubes are capped, and on the end facing out is a funnel leading to a clear plastic cup. After three or four weeks in the warm room, the emerald ash borers emerge. They are drawn to the light and collect in the plastic cups. We wandered among the kegs, peering at the prisoners. The beetles moved like little wind-up toys. They were a green so lustrous it was almost oily. The beetles were breeding stock. Their children would become wasp food.

At the time of my visit, the APHIS wasp farmers were rearing four species of Asian wasps: two species of *Spathius,* which lay their eggs on the beetle larvae; one species of *Tetrastichus,* which lay their eggs inside the beetle larvae, and one *Oobius*, which lay their eggs inside emerald ash borer eggs. Leah Bauer and Toby Petrice were right. The egg parasitoids were dwarfed by the *Tetrastichus* wasps we'd released into the woods. A flake of pepper was bigger.

Rearing such creatures is both a science and an art, Ben Slager said. "When you're dealing with living things, it's always finicky." Sometimes the ash logs get sodden and the emerald ash borer larvae turn to mush. Sometimes the wasps take ill. The wasp farmers have honed their craft over the last decade. "We've become far more efficient and utilitarian and smart," Sharon Lucik said. When the facility opened, in 2009, each wasp cost $50 or $60 to produce, Slager said, "something just ridiculous." That year, the facility reared about ten thousand wasps. Now each wasp costs between $1.50 and $2.50. In the two years before my visit, the facility had shipped more than a million wasps to dozens of states.

It has an easy logic to it. As we'd walked in the woods, Leah Bauer told me, "The only reason all our trees and plants are alive is because we have small parasitoid wasps and other natural enemies that are highly coevolved with their hosts." She gestured to a nearby hickory. There is an *Agrilus* beetle that is native to hickory, she said, and there are parasitoid wasps native to the beetle. The wasps spend their lives roaming a landscape of hickory trunks, searching for the beetle larvae, pupae, or eggs. The wasps depend on the beetle, the beetle depends on the tree, and the tree depends on the wasps. All the pieces fit together perfectly.

What could go wrong? This is perhaps a question better asked in the negative: What couldn't go wrong? Recall the two carnivorous snails Charles Elton mentioned in *The Ecology of Invasions*, introduced into Hawaii as counterpests in the mid-1950s: Neither succeeded in eliminating their intended mark, the giant African snail, but they caused the decline and possible extinction of numerous species of native Hawaiian snails. Between 1905 and 1929, authorities released more than forty-five counterpests of the gypsy moth into the eastern United States. Some of them attacked the gypsy moths, but they also attacked native moths, including the luna and Polyphemus moths that Étienne Léopold Trouvelot had tried to breed for silk. In the 1800s, people introduced mongooses from central and southwestern Asia into Hawaii, the West Indies, Mauritius, and Fiji. They were to be a counterpest of the rats infesting sugarcane fields. It didn't work. Rats are nocturnal, and mongooses are diurnal. The mongooses devoured rare creatures of all sorts, and drove at least one bird species to extinction. They sometimes come down with rabies.

Things have gone wrong, badly and often. These mistakes are still what the field is best known for, but Leah Bauer and other scientists I spoke with insisted that biological control is on the whole a safer, more regulated, more predictable activity than it used to be. Before the scientists in Michigan could release parasitoid wasps, they had to test whether they might attack other species besides the emerald ash borer.

The worst disasters of biological control have come from creatures like carnivorous snails and mongooses. These animals eat many different things. Scientists call them "generalists." Parasitoid wasps are pickier. They tend to lay their eggs on (or in) just a few species. Scientists call them "specialists." But the USDA scientists had to be sure. First, entomologists in China tried to get the wasps to attack the larvae and eggs of various moths and beetles, including two that feed on ash trees. The wasps ignored them. Next, in Massachusetts, Bauer's USDA colleague Juli Gould presented the wasps with nine of the emerald ash borer's sister species, members of genus *Agrilus*. In these "no-choice" tests, Gould gave the wasps only one option: Lay eggs either on the strange insect or nowhere at all. The wasps showed a deep commitment to the future of their race. Presented with no other option, they sometimes laid their eggs on or in the larvae and eggs of several other *Agrilus* species, among them the two-lined chestnut borer, *Agrilus bilineatus*, and the bronze birch borer, *Agrilus anxius*.

The question, then, is of risk and reward. In the post-mongoose, post–gypsy moth world, it's hard to gain approval to release counterpests of invasive plants. Many invasive plants are closely related to important crops. "That's really dicey," Bauer said. The potential for unintended consequences is too high. But if the unintended victim of a counterpest is another bug? "Most of the public doesn't really care if we kill native insects," she said. (She was careful to note that she personally doesn't feel that way.) The two-lined chestnut borers and bronze birch borers are nearly invisible. They appear only when their host trees are stressed. Most people would call them pests, if they called them anything at all. Who cares if the wasps attack them?

Dan Herms, for one. He's an entomologist and vice president of research at the Davey Tree Expert Company. He wrote his dissertation on the bronze birch borer. Back in the 1990s, he showed that it is deadly to European and Asian species of birches—just as deadly, it turned out, as the emerald ash borer is to North American ashes. The beetle is also, in its way, as full and rich a creature as a whale or a dog or a giant sequoia. It is the same shape as the emerald ash borer but the speckled bronze of spray-painted street performers. "They're beautiful," Herms said. He wonders whether they are doomed to fall victim to the counterpests. North America is home to at least 170 species of native *Agrilus* beetles. The U.S. Department of Agriculture vetted just nine. "The USDA says, 'Eh, well, it's a test,'" he said. "I think it's native biodiversity."

But it comes back to this: The bronze birch borers and two-lined chestnut borers are pests, invisible, unknown. The trees we see and know. They are the frame that held up a forest, providing food and shelter for arthropods and birds, mosses and mammals. They are street trees and yard trees and timber trees, the stuff of black ash baskets and baseball bats, tool handles and oars. By what system of value does the bronze birch borer carry equal weight? Herms used to get upset about it, but lately he's taken a more philosophical view. "They've done it," he said. "It can't be put back in the box."

There is another question, not just of whether biological control is worth the risk but whether it is even solving the right problem. There are two theories about the biological constraints that keep creatures like the emerald ash borer from running rampant in their native lands. The first theory is that the constraints come from a third party: Other creatures, like the parasitoid wasps, keep the emerald ash borer's numbers from soaring out of control in Asia. These creatures were not present in North America, so the borer was able to spread with impunity. This is the "enemy-release" hypothesis of invasions. It is the basis for the entire practice of biological control.

But the emerald ash borer's old enemies haven't yet stopped its spread in North America. In recent years, Leah Bauer and other entomologists have conducted surveys to see how many ash borers are killed by the parasitoid wasps they released. They estimate that the wasps killed something like 20 percent of the beetles in the areas they surveyed. To put it another way, so far the wasps are about as deadly to emerald ash borers as foxes are to rabbits, or owls are to mice—individually lethal, but not a species-wide threat. By contrast, researchers estimate that in southeastern Michigan, near the epicenter of the invasion, the emerald ash borer killed 99.7 percent of mature ash trees—all of them, give or take. It could only be described as a plague.

These are rare numbers. "As a rule, viruses, microbes, and parasites do not kill the majority of their victims," wrote Charles Mann in *1491*. The worst pandemic of modern times, the 1918 outbreak of Spanish flu, infected five hundred million people and killed fifty million, or roughly 3 percent of the world's population. It didn't come close to what happened after Europeans arrived in the Western Hemisphere. The human invaders brought with them measles, tuberculosis, smallpox, typhus, influenza, yellow fever, and other diseases. Researchers have estimated that in the several centuries after Columbus sailed into the West Indies, these diseases killed as much as 95 percent of the Native American population. "The epidemics killed about one out of every five people on earth," Mann wrote. He quoted geographer W. George Lovell, who called it the "greatest destruction of lives in human history." Scientists chalk up the shocking death toll to a lack of shared history. Europeans and Asians had lived for centuries with these diseases, and with the livestock that harbored them. Native Americans had not. Their immune systems were well tuned to fighting the diseases they had encountered before Europeans arrived, but not to the new Eurasian diseases.

This is the second theory of invasions, the "naïve host" hypothesis. As Peter Brannen wrote in *The Ends of the World*, researchers think that evolutionary naïveté might have been part of what doomed much of the world's megafauna. Recall the fox of the Falklands, lured with one hand and stabbed with the other. Maybe the mastodons and ground sloths of North America were the same, wary of wolves and saber-toothed cats but heedless of a new, more dangerous enemy. The elephants and rhinos of Africa and Eurasia, meanwhile, had lived alongside hairless apes for millions of years, and fared better. The story of the emerald ash borer and the ash trees of North America could be explained the same way. By this view, it wasn't wasps that constrained the emerald ash borer in its native Asia but the ash trees themselves. The Asian trees had defenses, honed by millions of years of shared history. The North American trees lacked these defenses.

There is evidence in favor of this explanation. Ecologist Chad Rigsby led a series of recent studies that compared the defenses of various species of North American ashes with the Manchurian ash, a native host of the emerald ash borer. Rigsby and his colleagues found that an enzyme that causes organic compounds to oxidize works much faster in Manchurian ash than it does in North American ashes. This oxidation process is what causes apple slices to brown. When an ash borer larvae bites into the trees' cells, the various compounds mix and begin to oxidize in its gut. "It's like when you drop a spoon down the disposal in your sink," Rigsby said. "It's wreaking havoc and creating all kinds of problems for the insect." Sometimes the ash borers kill a Manchurian ash, but often a Manchurian ash kills the ash borers.

The most persuasive argument for the naïve host hypothesis is the fate of North American ash trees planted in China. Chinese foresters and horticulturists had experimented with various North American ash species beginning in the 1950s and '60s. In Houping Liu's report on his 2003 expedition to look for counterpests of the emerald ash borer, he wrote that North American white and green ashes were so susceptible to the emerald ash borer that foresters in China had stopped planting them. The North American ashes couldn't survive the emerald ash borer even when they were surrounded by the full suite of the emerald ash borer's native enemies.

Rigsby considered this unsurprising. Burrowing around under the bark, in the cambium, the ash borer larvae attack the most sensitive part of the tree, Rigsby said, where it grows and distributes nutrients and water. The trees would face strong evolutionary pressure to defend those tissues, he said. Trees that can fight off the beetle survive, and trees that can't, don't. Any genes that provide a defense would have long ago spread through the Manchurian ash population. "There's really nothing that a biological control agent can do to rescue a tree when that tissue is being damaged," Rigsby said. "It's an absolute no-brainer, in my mind, that this is never going to work."

Many of the people I've spoken with during my research are skeptical of biological control. Is it worth the risk to nontarget species? Is it worth the money? Does it even work? But the entomologists carrying out the emerald ash borer biocontrol efforts argued that it was at least a chance at saving what remains of North America's ash forest. "One extreme view is 'Let's not do anything,'" said Jian Duan, one of the USDA entomologists I met in Washington, D.C. "'Just let some of the ash die, and eventually maybe the system will recover.' So what, wait it out a thousand years?"

I asked Leah Bauer whether she thought the wasps were succeeding. "I can't say," she said. "I think it'll be beyond my lifetime to know if emerald ash borer biocontrol actually works." Bauer told me that the most effective natural enemy of the emerald ash borer in North America is still woodpeckers. Researchers have discovered the parasitoid wasps attacking bronze birch borers.

In an 1878 speech, German doctor and naturalist Heinrich Anton de Bary used the term "symbiosis" to describe the most intimate of interspecies connections, those in which two or more species live together in physical contact. Sometimes these relationships are of mutual benefit. Lichen, for instance, is not a single creature but a seamless alloy of algae and one or more species of fungus. Most terrestrial plants are part of a similar partnership, joining their roots with mycorrhizal fungi. These fungi help the plants absorb nutrients and water, and the plants provide the fungi with sugars. Humans, too, are mutualistic symbionts. We are covered in and filled with great numbers of microorganisms of countless species, some of them important to our health. We are told they like yogurt.

In other symbioses, the association is beneficial to one party and is simply tolerated by the other, which gleans no obvious benefit. These creatures, said de Bary, "feed from the crumbs that fall from the table of the rich." Finally, there are parasitic symbioses, where one organism gains at the other's expense. Even Charles Darwin, great admirer of the world's diverse life-forms, found some of these parasites distasteful. In a letter to Asa Gray, he wrote, speaking of a parasitoid wasp, "I cannot persuade myself that a beneficent and omnipotent God would have designedly created the Ichneumonidae with the express intention of their feeding within the living bodies of caterpillars." But perhaps Darwin simply hadn't gone deep enough to see the toolmarks of divinity.

In a 1961 paper, British zoologist Richard Askew charted the relationships of the dozens of wasp species that create and inhabit oak galls in the British Isles. Oak galls are full of action. Askew depicted the arrangements in a series of flowcharts. The charts look like alluvial fans. In one of them, the first wasp lays its eggs in oak leaves, inflaming the leaf tissue and creating a gall, a round growth of about a tenth of an inch in diameter. Another wasp parasitizes the first wasp. A third wasp parasitizes the second. A fourth wasp parasitizes the third. There's no compelling reason to think such stacks don't go even further, wasps all the way down.

Mike Gates, the USDA taxonomist I met at the Smithsonian's entomology department in Washington, D.C., showed me a slide containing one of the world's smallest known multicellular organisms, another parasitoid wasp. The slide looked empty. For visual aid, he produced a photograph taken with an electron microscope. It showed two blobs and a wasp of equal size. The blobs were an amoeba and a paramecium, each a single cell. The wasp was fully formed, with legs and compound eyes and even little hairs. It floats through the ether and lays its eggs in the eggs of thrips.

Think of the timing involved. Think of the physics. I stood there looking at the photo with my mouth open; maybe a tiny wasp flew in. It seemed so improbably perfect, so finely tailored. How does such a way of life come about? One answer is that such symbiosis is the product of long coevolution. Due to their minute differences, individuals of a species are more or less successful in passing on their genes. Genes that contribute to success accumulate in a population. In a situation of coevolution, the success or failure of each species also depends on the success or failure of the other, so over time the tools of one or both become more and more specialized. In this way, the parasitoid wasp comes to look a bit like an apple corer or an avocado scoop: a tool well suited to its job but likely useless for any other.

But there is another explanation for how these relationships could arise. In 1985, ecologist Daniel Janzen proposed what he called "ecological fitting." He'd been working in Costa Rica's Santa Rosa National Park, home to some 650 species of plants, 3,100 species of butterflies and moths, 200-plus species of seed-eating beetles, 58 species of mammals, 250 species of birds, and many others. The park was "quite literally crawling with complex biotic interactions," he wrote. Many of the species seemed built for one another. But nearly all of the park's species had large geographic ranges, he wrote. "Over their wide ranges, these species interact in many complicated ways with many species that do not occur at Santa Rosa." This suggested that not all of the interactions Janzen saw at Santa Rosa were the product of coevolution but were, rather, the result of biogeographical happenstance. "A species does not have to evolve in a habitat in order to participate in the interactions in that habitat," he wrote. Species bring what tools they already have and make them work.

The fossil record, too, suggests that the way species arrange themselves together is flexible. It is full of "no-analog assemblages," species mixed together in ways unlike any that exist today. Sometimes these mixes seem contradictory. Today spruce trees and ash trees barely overlap in eastern North America. They suggest different places, different conditions. But just fourteen thousand years ago, at the end of the Pleistocene, spruces and ashes coexisted and were among the region's commonest types of trees. One way to understand this type of no-analog assemblage—and ecological fitting more generally—is through the lens of the Hutchinsonian niche. A species may be capable of surviving in conditions unlike those we currently observe it under, even in conditions that don't currently exist anywhere on Earth. If conditions are right, as they once were, ash trees and spruce trees will grow side by side. "The current state of the Earth system, and its constituent ecosystems, is just one of many possible states," wrote paleoecologists John Williams and Stephen Jackson in a 2007 paper on no-analog ecosystems and niches, "and both past and future systems may differ fundamentally from the present."

In recent decades a cadre of scientists and journalists have pointed to this ecological flexibility to argue against what they see as invasion biologists' undue insistence on keeping species in their places. One of invasion biology's central claims is that the knitting together of New Pangaea will lead to an overall loss in the number of species. Not so, these reformers argue: Carrying things to new places will in fact lead to more species. This argument makes sense on its face. In life, separation is what drives creation. One population of a species is isolated from another, either in space or in habit, and eventually becomes something different. Maybe it begins with ecological fitting. Suppose that North America's ash trees disappear and that most emerald ash borers disappear with them. The counterpest wasps that have begun to attack bronze birch borers might begin to have more reproductive success, on average, than the ones that attack only emerald ash borers. Given enough time, perhaps one species of wasp becomes two.

Indeed, we know that this process has happened many times in Earth's history. At the end of the Permian period, about 252 million years ago, more than 90 percent of species went extinct. Life recovered. The end-Permian mass extinction was just one of five such events in the history of life on Earth. The most recent came at the end of the Cretaceous period, 66 million years ago. It was caused by an asteroid that struck the Yucatán Peninsula, possibly in conjunction with widespread volcanism. The non-avian dinosaurs were among the life-forms that disappeared. This time, when life regained its diversity, it was with mammals as the dominant land animals. "Lots of bad stuff has happened on this planet," paleontologist Alycia Stigall told me. "Life bounces back."

Stigall studies ancient invasions, particularly in marine ecosystems. In the late Devonian period, which began 419 million years ago and ended 358 million years ago, the planet suffered a biodiversity crisis, in which the rate of extinction stayed relatively steady but the rate of new species forming decreased. What seems to have happened is that the sea level rose, she said, allowing long-separated marine species to meet. For a time, the number of species increased in many places, as generalist species spread. But over time, the absolute number of species fell. Specialists disappeared. Generalists were able to mix and remix with others of their kind, and so, without separation, the creation of new species slowed to a trickle. This is the scenario that some imagine as the endpoint of the New Pangaea. In 1999, entomologist Michael Samways coined the term "Homogenocene" to describe a time when the whole world is the same. Writer David Quammen called it the "planet of weeds."

In his 2016 book *Inheritors of the Earth*, British ecologist Chris Thomas cast this scene in a positive light. Looking 10 million years into the future, he predicted that the most common creatures of that time will appear remarkably similar to the most common creatures of ours. "The heirs to the world are not bizarre, weird things and ancient evolutionary relics—those will be the species that are dying out," he wrote. "The success stories are already all around us. Look out of your window and the chances are you will be staring at the future." It has an almost triumphant tone: Forget the rhinos and sequoias, the strange and the rare, the things conservation was originally about.

When I asked her about this spin on things, Stigall agreed that many of the specialists of today are likely doomed. Many scientists think we have entered a new era of mass extinction; the time is the Anthropocene, the place is New Pangaea, and the event is the Sixth Extinction. In a study published in 2004, Chris Thomas led a group of researchers in assessing the extinction risk in a set of sample regions that contained roughly a fifth of the land on Earth. Using the worst-case scenarios of climate change, they projected that as many as 37 percent of the species in those areas would be "committed to extinction" by 2050. Thomas's book, by this light, comes to look more like a statement of radical acceptance: By a long enough view, even mass extinction fades to insignificance. Life will continue, and new specialists will appear. But humans won't be around to see it, Stigall told me. "We're not going to get to the point where it rebounds."

We may hope that time will eventually undo the damage we're doing to life's diversity. But we don't need to look to the distant future for stories of resurrection. What is now happening to ash trees once happened to the eastern hemlock. Around 4,800 years ago, the amount of hemlock pollen suddenly plummeted across North America. Within 50 years, the quantity of the tree's pollen fell by 90 percent. "I postulate that a sudden outbreak of hemlock loopers or similar pests caused the hemlock decline," wrote paleoecologist Margaret Davis in a 1981 paper. Hemlock loopers are moths whose caterpillars eat hemlock needles. Maybe the loopers had somehow escaped the other organisms that once held them in check, Davis wrote, or maybe the eastern hemlocks were evolutionarily naïve.

But after a thousand years of near-total absence from the forest, the eastern hemlocks began to recover. After another thousand years, they'd returned to their prior abundance. Perhaps the loopers' natural enemies caught up to them, Davis wrote, or, more likely, the trees' brush with extinction left behind only trees that could resist the loopers' attack. Over time, genes that conferred resistance spread through the population. The hemlocks evolved. This is a plausible explanation for what happened, anyway, and it suggests another route to fixing the damage done to forests by new invaders.

On a sunny morning a couple weeks after my visit to Michigan, I was out walking in the woods near the Connecticut River with a man named Christian Marks when we came upon an enormous American elm. It was huge—four feet in diameter, more than a hundred feet tall. Staring up at it, I realized with surprise that many of the woods of eastern North America were once filled with similar trees. The forest around us seemed to shrink by comparison.

Marks is an ecologist with the Nature Conservancy. Looking for enormous elms in the woods near the Connecticut River is one of the things he does for a living, so at first I thought he'd led me to this tree on purpose. But he was as surprised as I was. "I must have driven past it at least a dozen times," he said, looking up at it. He turned to me. "Do you have a GPS app on your phone?" His phone was out of battery, and he wanted to be sure he knew where the tree was, so he could come back and take cuttings.

In theory, the solution to the problem of naïve hosts is simple. It's the same as the solution to the problem of invaders that have escaped their natural enemies: Just introduce the missing constraints. It's been nearly a century since Dutch elm disease arrived in the United States. Any elm still standing has likely faced the disease and survived. Instead of waiting for millennia in the hope that the survivor elms' genes will spread through the population, humans might facilitate this transfer.

This is another old idea. Almost as soon as chestnut blight began sweeping through the forests of Appalachia, scientists began trying to breed blight-resistant American chestnuts. But as Susan Freinkel recounts in *American Chestnut*, this turned out to be more of a tree-length task than a human-length one. Botanists first tried crossing American chestnuts with Asian chestnuts. As sometimes happens, though, two beautiful parents made an ugly child. Instead of massive and stately, the trees were bushy. The half-and-half hybrids also wound up only about half-and-half blight resistant. Others died of cold that their American sires would have survived. Decades went by. Finally, in the 1980s, scientists began planning a subtler kind of hybrid, using a technique called backcrossing. They would first cross an American chestnut with a Chinese chestnut, the most blight-resistant member of the genus. They would then breed the best of this first generation with another American chestnut, producing a hybrid that was three-quarters American chestnut, one-quarter Chinese chestnut. They would repeat this process again and again, until they wound up with a tree that was fifteen-sixteenths American chestnut. It would have all the physical characteristics of an American tree and the blight resistance of an Asian tree.

In the 1950s, scientists began a parallel program to breed American elm trees resistant to Dutch elm disease. This program focused on survivor trees. Researchers took cuttings from trees that seemed to have survived the disease, then rooted and grew the cuttings. When the saplings got big enough, scientists injected them with the fungus. But the survivor tree program started too early, Christian Marks told me. Not all of the vulnerable trees were dead, which made it hard to sort trees with immunity from trees that simply hadn't yet died. "They were just looking at random," he said. After testing a hundred thousand trees, Marks said, researchers found only nine that could resist the disease.

Clones of these nine survivor trees are now available to the public, bearing trade names both aspirational and nostalgic: New Harmony, Valley Forge, Princeton, Jefferson, Lewis & Clark. But all of the cultivars "diverge in one way or another from the ideal," said Bruce Carley, a horticulturist in Acton, Massachusetts. The classic American elm had a shape like an incandescent lightbulb, with narrow hips and wide shoulders. So far the cultivars all fall short, Carley told me. Some have the wrong shape—more ball hitch than lightbulb—while others grow too slowly, are sensitive to cold, or break apart under the weight of their own crown. Most can be relied upon to resist Dutch elm disease only some of the time.

But Marks thinks these problems could be solved in the coming decades. Finding resistant trees is getting easier. Time has separated out the chaff. Now when Marks finds a large elm, there's a good chance it has not merely avoided Dutch elm disease but survived it. He sends the cuttings of elms he finds to Jim Slavicek, a Forest Service researcher in Delaware, Ohio, who will test which trees are truly resistant.

When the elms died and left city streets bare, people often planted ashes to replace them. In forests, it was often ashes that filled the holes the dead elms left in the canopy. Now the process of resuscitation begins again, this time for ash trees. This project is more difficult than the quest to revive the American chestnut and the American elm, since it involves not one but sixteen species of North American ashes, only some of which can hybridize with east Asian ashes. Scientists have begun the search for ash trees that can resist the borers. They mine low-grade ore: In 2016, geneticists Jennifer Koch and Jeanne Romero-Severson estimated that just one-half of 1 percent of green ash trees, *Fraxinus pennsylvanica,* show any resistance to the beetle larvae. Even fewer seem to be fully resistant. But maybe this is enough. Koch and Romero-Severson wrote that lingering ashes seem to use several different mechanisms to defend themselves. Perhaps these trees could be bred together, stacking their defenses. With the help of counterpest wasps, Koch and Romero-

Severson wrote, maybe such an ash could survive in the wild. So begins the work of decades.

There are potential shortcuts. In recent decades, scientists have developed techniques to directly edit the genome. With these techniques, useful genes can be not only made more prominent within a species but also swapped between species. The first genetically engineered crop was planted in the early 1980s. As of 2014, more than 90 percent of the acreage of soybeans, corn, and cotton in the United States was genetically engineered. Some of these crops were engineered for resistance to bacteria, viruses, fungi, or nematodes; some to tolerate drought, herbicides, or insecticides; some for better taste or for nutritional richness. So far, such editing is rare for forest trees, and as a conservation method, it is entirely absent. But recently some have suggested it.

In March 2018, the Atlanta Botanical Garden and Jason Smith, the plant pathologist at the University of Florida, Gainesville, who helped identify the *Fusarium* fungus that was plaguing torreyas, put on a conference they called the Torreya Tree of Life, held at Torreya State Park, in Florida. Many of the people leading official efforts to save the Florida torreya attended. The keynote speaker was the eminent ecologist E. O. Wilson. At the end of March, *Yale Environment* 360 published author Janet Marinelli's report on the conference. Attendees agreed that the Florida torreya's situation was an emergency, she wrote. More of the trees were dying each year. Fewer than a thousand remained in the wild, and these were little more than sickly sprouts. The disease was winning. It seemed the only hope for the tree to have a future in the wild would be for people to tip the odds back in the tree's favor. But traditional methods of breeding for resistance to the fungus would take too long, Marinelli wrote. Jason Smith wanted to use a new gene-editing tool called CRISPR (pronounced like the lettuce drawer) to genetically modify the tree. In a press release reprinted in the *Tallahassee Democrat*, Smith said, "This isn't typically how conservation is done, so we're excited to be trying this novel approach."

The news upset Connie Barlow. To her, it was a travesty, an abomination, a sin, a waste of money. What were Smith's motives? It seemed to her like a sick hypocrisy that people who were supposedly dedicated to saving the tree would agree, with little apparent debate, to alter its very genome, meanwhile ignoring the cheaper, more elegant solution Barlow had been offering the whole time: Move the trees north. "You don't need to be a scientist," she said. "You just have to have a logical brain." She vowed to fight any effort to encrispen the Florida torreya.

"That's perfectly understandable that somebody might be concerned," Smith told me. It was easy to imagine how things might go wrong. Maybe you spend the time and money to build a perfect defense, then the fungus evolves a little and you have to start over. Or you could fix one problem and the tree becomes less fit in other ways. Maybe you've changed how the tree interacts with other species. But Smith said the edits he's proposing would be relatively minor. Researchers would identify parts of the torreya's genome that defend against the *Fusarium* fungus, however inadequately, and would then alter the genes so they were permanently switched on, like a broken gumball machine, continually spitting out proteins. "It sounds kind of scary, gene edits," Smith said. But people do the same thing all the time with less intention. When someone crosses one species of tree with another, they're editing the entire genome, he said. The genome even changes during an organism's lifetime, responding to stimuli of all kinds. When you take a drag on a cigarette, you're editing your genome, he said. Have a drink, and you've edited your genome. It begins to sound almost pleasant.

Assisted migration or assisted evolution? For now, both of its would-be saviors are stuck. By 2018, Barlow's Torreya Guardians had sent thousands of Florida torreya seeds to volunteers across the eastern United States. But in recent years, they'd had trouble obtaining seeds. The arboretums and botanical gardens they relied on had mostly stopped supplying them. Barlow had embarked on a mission to force the state and federal agencies responsible for overseeing the torreya's recovery to account for the seeds that the trees in its care produced, which, in her view, were being heinously wasted. "You'd rather have these seeds go to the squirrels," she said, speaking of these agencies, "than have them go to us for free-planting seed experiments!"

Jason Smith, meanwhile, has no funding. "Honestly, if we can't get some more financial support, I don't know how much I'll be able to do," he told me. He has applied for various grants, but it's difficult. To get natural resource grants, he said, "you have to show that the species either has economic importance or affects a broad geographical range." Chestnuts, elms, and ashes all fit the bill, but Florida torreya is geographically rare, economically irrelevant. Similarly, funding for endangered species tends to focus on certain types of places. "If you're working on an endangered species on a military base, there's money," Smith said. "Well, torreya doesn't occur on military bases. There's another chunk available for federal lands. Well, torreya doesn't occur on federal lands. There's another chunk of money for endangered species in the Everglades. Torreya doesn't occur in the Everglades. See what I mean?" The funding gene does not express.

A year after my visit to Michigan, I hiked through a clear-cut with Kevin Evans and Samantha Slingerland. We were in Dartmouth's Second College Grant, a twenty-seven-thousand-acre swath of land in northern New Hampshire, abutting Maine. The state gave Dartmouth College the land in 1807, as a means to pay for itself after it blew through the First College Grant. Dartmouth intended to sell the land to settlers, as it had with the first, but the second is cold and far-flung, and the settlers weren't buying. Instead, the college sold off timber, a little every year.

Timber is no longer an important part of Dartmouth's funding, but the grant remains a working forest, now the oldest in the country. Evans is its director, Slingerland his assistant. The grant's foresters, mostly free from worry about quarterly profits, focus instead on practicing an ecologically sound type of logging. They allow trees to grow old and fat and, when the time finally comes, fell them one by one. The phrases foresters use to describe this type of harvesting are "long rotation" and "selection cutting."

The clear-cut, that is, was unusual. It was a quarter acre, a rough square about one hundred feet to a side. Just a few scattered trees remained standing. The small trees the foresters couldn't use for lumber were laid flat like cut hay. These trunks help prevent runoff while the forest grows back in, Evans explained. But they made walking difficult. The fallers had cut the trees off in punji stake points. As we wended our way over the jagged, wobbling logs, Evans pointed out surveying flags of various colors. They marked where, just a few days earlier, foresters had replanted the cut. Purple meant red spruce. Dark green meant hemlock. Light blue meant birch. Yellow, pignut hickory; light green, white pine; dark blue, black cherry; brown, oak; and pink, aspen. The natural forest that surrounded the clear-cut was mostly birch, beech, and maple. "We call it the 'BBM forest,'" Evans said. The forest of seedlings they'd planted, though, more closely resembled the forest far to the south.

In 2007, ecologists Connie Millar, Scott Stephens, and Nate Stephenson (my guide in Sequoia National Park) published a paper that laid out three broad strategies foresters might employ in the face of climate change. The first option was to try to make forests more resistant to change. They compared this to "paddling upstream" but wrote that in some forests, "maintaining the status quo for a short time may be the only or best option." The second option was to try to make forests more resilient, better able to return to their prior state after a fire, bark beetle outbreak, or other such disturbance. This, too, would become more difficult over time, Millar, Stephens, and Stephenson wrote. The third option was to help forests change. Foresters might swap one species for another that might be less vulnerable to global warming or an invasive enemy. They might look for unusual patches of acidic soil or alkaline soil where they might seed the refugia of the future. Or they might establish neo-native forests, returning trees to places where they'd lived in the recent geological past, as Millar had suggested for Monterey pine a decade before.

In 2014, a network of scientists and land managers began an international experiment to test the three options Millar, Stephens, and Stephenson had laid out. In forests scattered across the United States and Canada, they set up a series of plots where they will attempt to help the forest resist climate change, be more resilient to it, or transition to something new. The Second College Grant is one of the land managers partnering on the project. The clear-cut plot I visited with Kevin Evans and Samantha Slingerland was a test of the third option, an experiment in transitioning the forest to something different. The mix of species they'd planted was more typical of the forests a hundred miles south, differing only in one key detail: They had included chestnuts.

Three decades after they began their backcrossing effort, scientists with the American Chestnut Foundation have created a tree that is one-sixteenth Manchurian chestnut and fifteen-sixteenths American chestnut. It is blight resistant (although not as blight resistant as Manchurian chestnuts), but otherwise resembles American chestnuts. "When I heard they were doing this, I said, 'That would be cool to include American chestnut,'" said Paul Schaberg, speaking of the Second College Grant experimental plots. Schaberg is a Forest Service plant physiologist. He worked with the American Chestnut Foundation to refine the hybrid, creating a tree that can tolerate not only the blight but also the cold of northern New England. The foundation gave the Second College Grant one thousand seeds. With the chestnuts, the experiment creates a forest that is not only relocated but novel in composition. American chestnuts weathered the Wisconsin glaciation in the southern Appalachians, and they spread slowly and steadily north throughout the Holocene. When chestnut blight arrived, they still hadn't made it this far north.

Orange flags meant chestnut. As we hiked across the ragged cut, I picked out every orange flag I could. I wanted to see a chestnut. They had been planted inside white plastic tubes, which were meant to warm the ground a little and protect the seeds from mice and deer. It has been a long time since any animal had a regular meal of American chestnut, but surely the taste for it remains.

Finally I found one. Even though I had never seen a mature chestnut, I recognized its leaves. They were serrated like the mouth of a carnivore. The tree's Latin name is *Castanea dentata.* The species name means "toothed." When I saw it, the tree was just two leaves poking up from the dirt, but if things go well, it could grow big someday. There are old pictures of people standing beside American chestnuts, dwarfed by them.

There is evidence to suggest that the chestnut forest was a human creation all along. Like giant sequoias, chestnuts tolerate fire and even benefit from it, given the right dose and timing. In the absence of fire, they will eventually be replaced by other trees. Paleobotanists Hazel and Paul Delcourt wrote that the eastern chestnut-oak forest of the southern Appalachians remained steady for some four thousand years, steadiness they thought suggested a human hand on the wheel. Native Americans were lighting fires, shaping the world to fit. The parklike forests filled with massive trees, according to this view, were reflections not of some inherent order but merely of human favor.

After the blight arrived, the trees became what scientists call "functionally extinct." They did not disappear completely, but those that remained were shrunken and damaged. It is similar to what happened to North American elms and Florida torreyas and to what is happening to North American ash trees. Something similar will undoubtedly happen to many more trees in the future. Some especially beloved species may earn the sustained attention that American chestnut has, but many probably will not. Even in cases where people can repair the tree, the ecosystem it was a part of will be gone, rearranged in its absence.

Still, looking down at the little chestnut, product of assisted evolution and assisted migration, I could understand why people had tried for so long to repair it. When I see the old pictures of people standing beside big chestnut trees, my chest aches and I miss something I never experienced. The trees were beautiful. Ecologist Deborah Rabinowitz once wrote, "Lost species hold a particular fascination, rather like ships lost at sea." Maybe the only thing stranger is to see them return.

6

THE FUTURE

The two defining characteristics of a tree—its height and its woodiness—are entwined. If a tree were not woody, it could not grow tall. The cell walls of all plants are built of a material called cellulose. Cellulose is strong under tension. Like an inflated balloon, a cell made of cellulose will pull taut when flush with water. When the plant made mostly of cellulose dries out, though, the tension is released and the plant wilts. The cell walls of woody plants are also made of cellulose, but they're fortified with a substance called lignin. Lignin is strong under compression. A lignified plant cell is like a Lego brick. Stacked together, these cells can defy gravity. The resulting material, wood, is mundane in purpose but miraculous in practice. "If all the greatest aesthetes and engineers that ever lived were assembled in some heavenly workshop and commissioned to devise a material with the strength, versatility, and beauty of wood I believe they would fall far short," wrote Colin Tudge in *The Tree*. "Wood is one of the wonders of the universe."

Some wood is so soft that a person might pinch it between two fingers and leave prints. Other wood dulls saw teeth and drill bits. Some woods are strong, others light, others flexible. There is fragrant wood and resonant wood and wood that's best for grinding up and digesting into paper pulp. Some wood, it seems, is built to last forever. Walking among the bristlecone pines in California's White Mountains, I noticed that many of the trees were posed like dancers missing their partners. Thousands of years ago, they'd grown over rocks long since eroded away.

Civilization sits on wooden joists. Various anthropologists have pointed out that the Stone Age could as easily be called the Wood Age, because that's what people used to build most of their shelters and fires and tools. But in fairness to trees, Tudge wrote, "every age has been a Wood Age." Stone, bronze, and iron are useful materials, age-defining even, but in utility and ubiquity, they've never had anything on wood.

People have long worried that this wonder material might run out. As geographer Michael Williams wrote in *Deforesting the Earth,* northern China experienced a timber famine in the twelfth century, and Rome suffered its own a millennium before that. As human populations grew, what had typically been a local problem sometimes became a regional one. By the early 1600s, people across Europe were worried about a shortage. "The spoil and wasting of this necessary material is no less than a public calamity," wrote British gardener and writer John Evelyn in 1662.

In the New World, Europeans found what appeared to be an inexhaustible forest. In *Changes in the Land,* William Cronon quoted a New England colonist named Francis Higginson, who in the early 1600s wrote, "A poor servant here that is to possesse but 50 Acres of land, may afford to give more wood for Timber and Fire . . . then many Noble men in *England* can afford to do." But by the early 1700s colonists had begun to complain that the most valuable trees were gone. In some areas timber of any kind had grown scarce. Over the next century these worries swelled into fears of a more widespread shortage. "We have now felled forest enough everywhere, in many districts far too much," George Perkins Marsh wrote in 1864, in *Man and Nature.* "The earth is fast becoming an unfit home for its noblest inhabitant."

In 1907, the newly minted U.S. Forest Service issued a report by William Hall titled *The Waning Hardwood Supply and the Appalachian Forests.* ("Hardwood" in the title refers to the wood of angiosperms. Wood from gymnosperms is called softwood, regardless of its actual texture.) Despite strong growth of industry as a whole, Hall wrote, the amount of hardwood harvested in the United States each year had fallen drastically since 1899. Oak was down. Poplar was down. Elm and ash were down. Casting his gaze over the former timber strongholds of the Lake States and the mid-Atlantic, he tallied estimates of how much hardwood still existed and how much was cut annually. He tracked the rise and fall in prices for various types of lumber. He noted how much hardwood was used each year in the construction of houses, railroad cars, farm implements, barrels, and wagons. Numbers in hand, he predicted that the country had a mere fifteen years of hardwood left. "The supply is getting short," he wrote, "and the end is coming into sight."

What does the future hold? Business strategist Kees van der Heijden wrote in a 1997 paper titled "Scenarios, Strategies and the Strategy Process" that most of what we know about the future is due to inertia—an object in motion tends to stay in motion, and an object at rest tends to stay at rest. The very near future, then, is to a large degree predictable. A tree today will likely be a tree tomorrow; a forest today will likely be a forest tomorrow. But over time, small changes accumulate. "The further out we look into the future," van der Heijden wrote, "the more uncertainty enters into our consideration." The very distant future is so unknowable as to defy speculation. Having journeyed ten million years ahead, the time-traveling botanist might write to a friend and say, "There are no Pines."

But between the very near future and the very distant one lies a stretch of time that is neither so close that inertia has made much inevitable nor so far that the range of potential outcomes stretches toward the infinite. It is the future we are most able to choose. Many of our disagreements in the present stem from divergent predictions about this stretch of time. What options are still available? What realities must we accept? Does the risk of inaction outweigh the risk of action? Divining the contents of this Goldlocksian stretch of time is a central scientific endeavor. The same way that scientists use fossils and tree rings and other proxies to reconstruct the past, they will build models—simplified versions of the world—as a way of peeking at an otherwise unknowable future.

William Hall's model of the American timber supply was of the simplest type. He plotted the points of the past, drew a line of best fit between them, and extended this line out into the unknown. The resulting prediction was wrong. "The forests of America have not been totally devastated, timber consumers have suffered no dire consequences, and there has been no great commercial disaster resulting from timber shortage," wrote Arthur Mather in a 1987 paper. Hall had underestimated how fast the country's timber stocks were growing and had not foreseen the shift to oil as a primary source of fuel. The future is often stubbornly nonlinear. Expected disasters fail to materialize, and unexpected disasters arrive suddenly, seeming obvious in hindsight.

Today, using models of far greater sophistication, scientists project that between 2030 and 2052, Earth's average surface temperature will have risen by roughly 2.7 degrees Fahrenheit over its preindustrial average. The increase will be greater over land, especially in alpine and polar regions. Under worst-case scenarios, they project that by the end of this century, Earth's average temperature will be higher still; some parts of the Arctic could warm by more than 15 degrees Fahrenheit. These projections bring with them a host of other possible effects: crop failures, the collapse of fisheries, drought, fire, coastal inundation. Such a level of warming would scramble ecosystems and send species racing to keep up with the climate that suits them. Forests would die in great patches.

It is possible that none of this will come to pass. It is possible to imagine scenarios—however unlikely—in which this disaster we see looming never appears, or takes some lesser form, scenarios in which the world's living things remain in an arrangement that suits us. But it is also possible to imagine scenarios in which a lack of wood is not the worst of our worries.

After my visit to Dartmouth's Second College Grant, where I'd seen the newly planted chestnuts, I continued north. Northern New Hampshire is clothed in what looks from the road to be an unbroken forest, although just off the highway you find holes. I spent the night in one of these, a clear-cut so recent that the knocked-down birches still held their leaves. I woke at dawn to loggers speeding past on the dirt road, and drove the last few miles to the Canadian border. Past customs, the road straightened and pitched forward. As I drove down the hill, the road signs turned from miles to kilometers, the language on the signs changed from English to French, and the land itself transformed, woods swinging open into pasture.

When French traders arrived in the St. Lawrence River valley in the early 1600s, it was furs they wanted. Trees provided fuel and lumber, but mostly the traders saw the endless forest as an impediment, or worse. To seventeenth-century Europeans, "forests were places of terrifying eeriness, awe, and horror," Michael Williams wrote. They were unknowable, the lairs of witches and demons. "Human progress had seemed to be viewed in some proportion to the amount of woodland cleared, or at least used," Williams wrote. This last point was surely the most important: A forest was space that could be better filled with more useful plants. The traders settled in, cleared the land, and sowed it with seeds of wheat and barley, thus beginning their part in what is likely humankind's biggest collective project—the clearing of the world's forests. Civilization is a sculpture in relief, carved out of the woods. In southern Quebec, the artist seemed to still be at work. Beside the farmhouses sat long rows of split firewood. Logs lay piled high in clearings and rumbled past on trucks.

My first stop was in Quebec City, at the research offices of the Ministère des Forêts, de la Faune et des Parcs (the provincial equivalent of the U.S. Forest Service), where I had an appointment with Louis Duchesne, a researcher there. Duchesne has reddish-blond hair and blue eyes. He wore black jeans with boots. His ball cap said, "Tree Ring Society." Although we'd e-mailed back and forth without translational mishap, he'd refused to talk with me on the phone. As we crossed the parking lot to his truck, he apologized for what turned out to be perfect English. I, in turn, apologized for my perfect lack of French.

We drove north out of town into rain-shrouded hills. As we climbed, the forest around us turned from birch and sugar maple to a solid wall of balsam fir. The firs were steeply conical, with branches in pagoda-roof layers that from a distance gave the forest the warp and weft of burlap. Sometimes I could see light through the forest where only a roadside strip of trees concealed a clear-cut.

The careless razing of Quebec's forests for farmland soon turned to more purposeful harvest. Europeans started felling the white pines of Newfoundland in the mid-1500s. The tall, straight pines made ideal ships' masts. As the scattered settlements swelled with new arrivals, the lumbering spread west through Nova Scotia and New Brunswick. By the 1700s the inhabitants of New France started cutting their own white pines. As Quebec's pines were depleted, its lumberjacks turned their attention to the province's other forests, first for lumber, then later as a source of pulp for paper.

In the prosperous decades that followed World War II, demand for wood products of all kinds soared. Canadian forestry boomed, and the number of paper mills in Quebec grew. "To many in industry, government, and communities dependent on the industry, steady increases in the production of lumber, plywood, pulp and paper had become the norm," wrote Ken Drushka in a history of Canada's forests. "Stability had become confused with steady expansion" (a mix-up by no means limited to Canadian forestry). But it wasn't clear whether Quebec's once-endless forests could actually support this expansion. By the 1960s, the province ostensibly followed a model of sustainable harvest, where the amount of timber felled was, on average, equal to or less than the amount of timber grown. The problem was, nobody knew exactly what the vast forests contained, much less how fast they were growing.

A forest is a complicated, restless thing, with trees dying and sprouting in every direction. A forester, trying to grasp what is happening, will most often start their investigation by staking out a plot. They use this small patch of trees as a key to the surrounding forest.

Louis Duchesne and I turned off the highway into the Forêt Montmorency, a research forest. In winter, Montmorency is a popular destination for snowmobilers and cross-country skiers. A sign in the visitor center advertised a nonlethal safari for *l'orignal,* the moose. Past the visitor center, we drove through a gate onto a dirt road, which Duchesne took at a speed I've come to associate with foresters driving company trucks. Finally we reached a clearing near a small lake. Duchesne donned a yellow raincoat. Then we headed down a path into the woods.

The air was thickly humid and smelled of a Christmas tree lot. The forest was mostly balsam firs—*sapins baumiers* in French—and a few scattered white spruces, *épinettes blanches.* We were in the wettest part of Quebec, Duchesne told me. The ground was sodden, carpeted with feathery mosses and draped with ferns. The trees' trunks were spattered with orange and gray and sea-green lichens.

We hiked up and down across the small hills that ringed the lake, then stopped, finally, in front of a balsam fir. It was indistinguishable from its neighbors—skinny and gray-barked, speckled with lichen and bare of limbs past head height—except that it was strangely dressed, with a skirt of metallic bubble wrap around its middle and, a few feet up, a sombrero of insulation foam held in place by wires. The skirt and hat were just rain gear, Duchesne explained. The important bit was in the middle. It was a small aluminum bar, one end bolted directly into the tree, the other end holding a little flat-ended piston that pressed against the bark. This was a dendrometer. As the tree swelled and shrank, it pushed against the piston. A wire ran from the dendrometer to a small box containing a computer, which recorded this movement. "You can see the heart of the tree, you know?" Duchesne said, pumping his fist in front of his chest. He pulled out a sheet of paper that showed this rhythm. Indeed, the zigzagging line looked like the one that blinks across a heart-rate monitor.

Around us were other devices: rain gauges and thermometers; instruments to measure humidity and solar radiation; a camera to track snowfall. There in the dark woods, the contraptions looked vaguely occult. I asked Duchesne if he'd seen *The Blair Witch Project*. He laughed politely. The array of machines record changes in local physical conditions, which researchers can then compare with the swelling and shrinking of the tree. The trees shrink during the day, when evaporation is high, Duchesne said. "Then, during the night, the trees drink." The dendrometer also records the tree's burst of summer growth, although the daily back-and-forth hides this trend line from easy view. On a drizzly, overcast day like today, he said, the trees grow more than usual.

For me, this was a new view. I'd thought about trees in various ways: as features of the landscape; as systems of pipes and solar arrays and chemical converters; as records of years and even millennia past; as habitat for other species; as sources of raw materials. Each version was illuminating in its own way, but my understanding of trees as living beings had always been from a distance in time, never reducible to the day-to-day existence I'm most familiar with. At that moment, though, looking at the pumping heart of a balsam fir, I could see that it was alive in a way unexpectedly similar to the way I was. I wondered, improbably, what it felt like when the stomata of its needles closed at night and it drew its trunk full of water after the thirst of the day. All around us were trees swelling and shrinking together, like a cloud of jellyfish dancing in time.

Each of the trees in the plot had a numbered dog tag, the same as the trees in the plot Nate Stephenson had taken me to visit in Sequoia National Park. As in that plot, foresters give the trees here regular checkups. With the help of scattered rain gauges and a gauge at the mouth of the stream flowing out of the lake, they can calculate exactly how much water enters and exits this watershed, and how much the trees are intercepting and pumping skyward. They can track the movement of nutrients through the system and can even—using a series of cameras—pinpoint the very day and hour of bud break, when the short growing season begins. It is a forest studied from top to bottom, down to the heartbeat of a single tree. I followed Duchesne down the path back to the truck, imagining the trees slowly pulsing around us.

The plot Louis Duchesne took me to see is just one of many thousands scattered across Quebec. In these plots, the province's foresters have counted and measured and cored the trees, identified slope and soil type, and noted whether the forest has suffered the effects of fire, insects, or logging. Back at the Forêt Montmorency visitor center, Duchesne showed me a map of these plots on his computer. As he turned on different layers of the map, the image of Quebec on his screen was speckled first with turquoise dots, then red, then blue. Each dot corresponded to a plot. The colors indicated the decade when foresters had surveyed the plot. As he added layers, the dots lay thicker and thicker on the map, until finally the southern half of the province disappeared under tricolored confetti. I was reminded of a line from a Jorge Luis Borges short story, "On Exactitude in Science": "In that Empire, the Art of Cartography attained such Perfection that the map of a single Province occupied the entirety of a City, and the map of the Empire, the entirety of a Province."

If the map of Quebec's forests were really the size of the province, of course, the clustered dots would almost disappear, space stretching out between them. The plots aren't meant to be comprehensive. Rather, they are a way to avoid comprehensiveness, a way of getting around the impossible task of measuring the entire forest, tree by tree.

Quebec's foresters started surveying scattered plots in the 1970s. At the same time, they began assembling a bird's-eye view of the province's forests, first with aerial photographs, then, later, with satellite imagery and lidar (a light-detection method similar to radar). To marry this broad image of the forest with the narrow, detailed view from the plots, they matched like with like. Analysts divided the aerial map they built into polygons based on similarity of forest type, soil type, tree height, elevation, slope, aspect, and other measures. Each of these polygons contained one or more of the plots that foresters had directly measured. What was known about the species composition, size, and growth rates of the trees in the plots was extended to the broader, unmeasured areas of the polygons they were a part of.

The result is a map of remarkable detail. The first layer shows only the rough vegetational territories: the temperate forest in the south; the boreal forest in the middle; and the tundra in the north. But as you add more layers, broad kingdoms give way to lesser fiefdoms: regions of sugar maple and bitternut hickory, of balsam fir and white birch, of spruce and moss, and many others. When you zoom in, the wide swaths of color signifying one species or another shatter into a mosaic of ceaseless variation. Along with forest type, the map shows the locations of past fires and insect outbreaks and logging, and where the forest was replanted. To an amazing extent, Quebec's vast forest is known.

Over forty years, Quebec's foresters have visited some four hundred thousand plots. Dozens of companies take part in the surveys, and an entire branch of government is devoted to overseeing the work. This effort is all to answer what seems like a simple question: Is the forest growing as fast as people are felling it? "The goal is to guarantee the forest is sustainable," Duchesne told me. "We have to inform the public of the state of the forest—'Your forest is going good.'" The *ministère* delivers the information it collects to the Office of the Chief Forester, which determines how much timber it will allow forestry companies to harvest. The process is not so different from what a wildlife manager might do, setting catch limits for fish or bag limits for game. *Forest in, forest out.*

But the amount of work involved in making that calculation helps explain why so little is known about forests in other parts of the world. There are few, if any, other regions where forests have been studied at such length and intensity. The oldest comprehensive satellite images of the world date to the early 1970s, when NASA launched its first Landsat satellite. New methods like lidar and increasingly capable satellites will continue to improve the bird's-eye view, but the scope of these data sets will always butt up against the recent invention of the technologies that enable them. Things are even spottier on the ground. There are few forests anywhere in the world that scientists have tracked for more than a century. In general, forest plots are more concentrated in North America and Europe, and more widely scattered in Asia, South America, and Africa. Bigger gaps between measured plots lead to more uncertainty about what the forests contain.

What all of this means, said Patrick Gonzalez, a forest ecologist at the University of California, Berkeley, is that we have a deeply incomplete picture not only of how forests in many parts of the world have changed in recent centuries but also of how they're changing now. Gonzalez led a recent effort to quantify the amount of carbon stored by terrestrial ecosystems in California. In 2006, the state government set a goal of reducing California's annual carbon emissions to their 1990 levels by 2020. It expected forests and other noncultivated ecosystems to absorb large amounts of carbon, offsetting a large part of the emissions. But nobody knew whether these ecosystems were actually absorbing more carbon than they were emitting. Gonzalez and four other researchers married the broad image of California's ecosystems provided by satellite data with the detailed, close-up information provided by a network of U.S. Forest Service plots. The places where foresters had measured tree by tree became the key to the broader areas where they hadn't. Gonzalez and his colleagues did this twice, building one carbon-storage map of the state's ecosystems as they existed in 2001 and another as they existed in 2010.

What they learned was disappointing, and perhaps an ominous sign of the future. "Regrettably," Gonzalez said, "we found that ecosystems were losing carbon in that period." A small fraction of the state's land had contributed a large amount of carbon loss. The culprit was fire. In a 2015 paper describing the mapping effort, the researchers wrote that if people did not reduce the use of fossil fuels, the problem would quickly get worse. Two years after the period covered in the study, the state fell into a long, hot drought.

Some things are predictable. One is the relationship between temperature and evaporation. Warmer temperatures cause higher rates of evaporation and, for plants, higher rates of transpiration from their leaves. This might allow plants to photosynthesize and grow faster, but it also means that they need more water. If temperatures increase but precipitation does not, plants will suffer water stress. In a 2009 paper, a group of researchers led by ecologist Phillip van Mantgem noted that temperatures in the western United States had increased by between one-half and three-quarters of a degree Fahrenheit in each decade since the 1970s. This, they wrote, could perhaps explain why foresters across western North America had observed trees dying at increasing rates, even under normal circumstances—that is, in the absence of drought or insect outbreaks. In long-term monitoring plots scattered from British Columbia to northern Arizona, they found that the rate of this "background mortality" had doubled, sometimes in as few as seventeen years.

Warmer temperatures will make droughts more damaging. I spoke to ecologist David Breshears, who has spent decades studying the effects of drought on the piñon and juniper forests of the American Southwest. In one experiment, Breshears and his colleagues uprooted small piñon pines and replanted them inside the Biosphere 2 glasshouses near Tucson, Arizona. They starved the trees of water to simulate the effects of drought. One group of trees they left at the ambient temperature, while in another group they raised the temperature by about eight degrees Fahrenheit. "Nobody had ever done that for an adult-sized tree of any species," Breshears said, "and yet that is essentially the question of climate change." The trees in the hotter drought died much faster than those in the cooler drought. The researchers repeated the experiment, this time with saplings, and raised the temperature more and more. "As it gets warmer and warmer and warmer, what happens?" Breshears said. "They died faster and faster and faster."

Warmer temperatures will also often benefit the insects that attack trees. Stressed trees make easier targets, and milder winters allow more insects to survive. Scientists have observed many insects that attack trees spreading poleward and upslope. Matt Ayres, an entomologist who studies southern pine beetles, which attack the pines of the American Southeast, told me the beetle is now more like a "mid-Atlantic beetle." Warmer temperatures could also give insects access to new species of trees.

Near the town of Saguenay, two hours north of Quebec City, ecologist Hubert Morin took me to look at a patch of forest that was infested with spruce budworms. Despite its name, the caterpillar's favorite food is balsam fir. The infested trees were grayed, shrouded in cobwebs. Demonstrating an admirable indifference to the mosquitoes swarming around us, Morin led me through the ailing forest. He stopped in front of a fir and, taking one of its boughs in his hand, rolled its needles between his fingers, revealing a mess of small brown caterpillars. Back in the 1980s, when Morin was a student, much of Quebec was in the grip of a budworm outbreak. When you walked through these forests, he said, the caterpillars would cover you. "It's not bad, because they don't eat people," he said, "but it is just a little bit disgusting." The falling frass—or caterpillar poop—"was like rain. It covered all the ground."

Currently the budworms rarely attack the black spruce, Quebec's most valuable timber tree. Black spruce's new growth emerges two weeks later in the spring than that of balsam fir and white spruce. By the time black spruce breaks bud, most of the caterpillars have already moved on to a different life stage and don't attack the tree's new growth. But warmer temperatures could lead the black spruce to bud out earlier, perhaps pushing it into the caterpillar's path. "We worry a lot about that," Morin told me. "Now it's not the caterpillar's preferred species. But what will happen in the future?"

Finally, warmer temperatures will likely bring more fire. Ever since there have been forests, they have burned. Each year, roughly 4 percent of the world's land burns—a number that is somewhat lower than in centuries past, as Stefan Doerr and Cristina Santín pointed out in a 2016 paper. "Global area burned appears to have overall declined over past decades," they wrote, "and there is increasing evidence that there is less fire in the global landscape today than centuries ago." This is largely due to people suppressing fires, they wrote. But this trend is unlikely to last. In many places, the fire season—that part of the year when fires are most likely to start and carry—is already much longer than it was in the twentieth century. Drought and insects pair with the effects of fire suppression to leave forests strewn with dead wood, ready to burn, as they are in California.

Just because we can predict that something might occur doesn't always make it less shocking when it happens. In the fall of 2017, a few weeks after a series of wildfires burned across California's Napa, Sonoma, and Mendocino Counties, I drove through the remnants of a neighborhood in a town called Glen Ellen, about an hour north of San Francisco. At the end of the flattened block, I met a man sifting through the remains of his house. Across the street, FEMA workers in yellow vests and masks wandered amid the ruins, spots of bright color in a scene of dun and black and gray. Charred washers and dryers poked from the ash like tombstones. The next fall it happened again. A fire swept through the town of Paradise, in northeastern California, killing at least eighty-five people. It was the deadliest wildfire in the United States in a century. For weeks my home in Oakland lay under thick smoke. The Sun was a sullen orange.

We can predict some of what a warmer world will bring: more stressed trees; hotter, deadlier drought; more insect outbreaks; more fire. But what will this do to the forests themselves? What will it mean for the arrangement of the world's living things? These effects are harder to foresee.

Between my trip with Louis Duchesne to the balsam fir forest of Montmorency and my trip with Hubert Morin to see the spruce budworm outbreak outside of Saguenay, I went on a field trip with a group of soil scientists to Parc National des Grands-Jardins, in the Charlevoix region, a couple hours northeast of Quebec City. The field trip was the scientists' final outing after a week-long soils conference. I sat near the front of the bus beside our guide, Serge Payette, a biologist at Laval University, Quebec. I'd called Payette earlier in the week to let him know I was in Quebec, and he'd invited me along on his turn as guest lecturer in what I assumed to be a two-birds, one-stone situation. He had long, gray-white hair swept back from his forehead and wore smoky oval glasses and a light blue button-up shirt over a second, lighter blue button-up shirt. That morning, as we'd stood next to a gas station, waiting for the bus to arrive, Payette told me that he'd started his research on Quebec's boreal forest forty years before, farther north, trying to more efficiently grow a local strain of *bleuets*—blueberries—that had proved resistant to oversight. He seemed to me improbably suave; with a scant few words in French to our driver, he sent the whole front of the bus into uproarious laughter.

We drove north. Ahead of us, mountains stretched out in a vast amphitheater, steep bare walls jutting from the forest. The geomorphologist sitting behind me explained that the mountains were ripples tossed up when a rock hit Earth some 450 million years ago. Payette noted that the towering faces were now a popular destination for rock climbers.

As we climbed into the mountains, Payette held a microphone and turned to face the scientists, naming the trees—sugar maple, ash, elm, hickory, yellow birch ("it's a nice-growing tree, okay"), balsam fir ("this is the best for Christmas trees"), white spruce ("I like very much white spruce")—and pointing out a favorite local eatery ("I know all the places around here for the best French fries"). We passed through a small village that Payette said was typically Quebecois, with small houses clustered close to the road—less snow to shovel from in front of the door, he explained—and a church with a tall, skinny steeple. We climbed toward the rim of the crater; then abruptly the forest changed, the loose weave of balsam fir replaced by dark shag. "All of the trees here are black spruce," Payette told the soil scientists. "I've worked a lot in black spruce. It's one of my favorite species." It was the far end of the forest that stretched three thousand miles to Alaska. What Payette wanted to show us was where this great boreal sweater had begun to unravel.

Suddenly, in a matter of feet, the dense black spruce gave way to something new. There were still black spruces, but now they were widely scattered, underlaid not by the usual thick moss but by a pale yellow-green lichen, which looked from a distance very much like a dessert my mom used to make from lime Jell-O and cottage cheese. The bus stopped, and Payette led us down a path to a park table, which he climbed onto, the better to be seen and heard. We gathered around him, looking across a landscape of rolling hills. In the distance was a fringe of the dense black spruce forest, what Payette called the closed-crown forest. The foreground was this second type of forest, what he called the lichen woodland. The lichen woodland is the characteristic forest of northern Quebec, he said, a transition zone between the closed-crown boreal and the tundra. But we were far south of that transition. This patch of lichen woodland was hundreds of miles out of place.

It started with a fire in 1999, Payette said. This itself wasn't unusual—as in California's Sierra Nevada, fire is a fact of life in the North American boreal. Unlike a giant sequoia, which weathers the flames behind its thick bark, the black spruce's strategy is to burn. The whole forest burns, hot and fast. But soon after the fire, the ground thickens with black spruce seedlings. Before too long there is a new black spruce forest, as thick and dark as what had been there before.

This, at least, is what has usually happened. Instead, here in the Parc National des Grands-Jardins, the forest took an unexpected turn, changing into something different. We could see both the old forest and the new, one superimposed on the other: the charred, densely spaced logs of the closed-crown forest and their more widely scattered progeny, poking from ground now stubbled with lichen. This patch won't go back to closed-canopy forest, at least not for a very long time, Payette said. "It's a change in dynamics. Come back in fifty years, it will still be an open woodland." He swept his hand across the lichened landscape. "The future is here."

Living things are arranged the way they are because of physical and climatic conditions, because of their interactions with other living things, and because of their past and present ability to reach otherwise suitable places. This arrangement is finely—albeit haphazardly—balanced, each part depending on many others. "Even a tiny incremental change in conditions can trigger a large shift in some systems if a critical threshold . . . is passed," wrote ecologist Marten Scheffer and limnologist (a scientist who studies lakes) Stephen Carpenter in a 2003 paper. "This phenomenon has many intuitive examples, such as the tipping over of an overloaded boat when too many people move to one side, the occurrence of earthquakes when tension builds up in the Earth's crust and the legendary straw that breaks the camel's back." The same thing, they wrote, occurs in ecosystems.

The fossil record is full of tipping points, Jill Johnstone told me. She's an ecologist who studies the boreal forest of Alaska and northwestern Canada. The transitions between Alaska's Holocene forests are good examples, she said. The early Holocene shift from deciduous forest to white spruce forest and the mid-Holocene shift to black spruce both happened fast, in a matter of centuries. "These patterns emerge so strongly in the paleoecological data," Johnstone said, "but we don't typically catch those dynamics happening."

Like Serge Payette, she's discovered places where the closed-canopy black spruce forest has begun to fray. In the Yukon Territory these patches have turned into a forest of mostly deciduous trees—birches, aspens, and cottonwoods. The changes in Yukon and Quebec seem to be driven by a similar set of events. In the Yukon, it was one fire, followed a few decades later by another. In Quebec, it was first an outbreak of spruce budworm, then fire. The trees that sprang up after the first disturbance hadn't had enough time to grow cones to replace themselves by the time of the second.

I asked Johnstone if she thought it was the beginning of the end of the black spruce boreal. "The threshold changes I've been describing, they're still unusual circumstances," she said. "But when we look at the conditions that give rise to those changes, we can ask ourselves if those are becoming more common." Sudden changes to small parts of the forest would eventually add up. Using Quebec's forest inventory data, Serge Payette found that, in the last fifty years, nearly 10 percent of the province's closed-crown black spruce forest had turned into lichen woodland. "If this trend is maintained," he wrote in a 2008 paper with François Girard and Réjean Gagnon, "closed-crown forests could disappear within about 550 years." More recent projections suggest that the North American boreal might change even faster. In 2019 a group of researchers led by Zelalem Mekonnen predicted that spruce forests would fall from dominance in Alaska by as early as the middle of this century and be replaced by deciduous forests. This change could bring a cascade of other effects: burning peat, thawing permafrost, expulsion of more carbon into the atmosphere, and more warming.

Many of the world's forests could transform with similar speed. In 2015, as the sequoias languished and tens of millions of the Sierra Nevada's other trees died, Nate Stephenson and Connie Millar, the U.S. Forest Service paleoecologist, published an article in the journal *Science*. They surveyed the forces affecting the world's forests (focusing in particular on those in temperate regions) and predicted that many of those forests would eventually hit tipping points: After a fire, drought, insect outbreak, or other disturbance, the forest wouldn't recover to its prior state. It would change into something different. They predicted that these changes would have far-reaching effects on human society, ranging from increased threat of wildfire to decreased water quality to flooding. Forests could turn from carbon sinks that dulled the effect of human carbon emissions to carbon sources.

Stephenson and Millar ended the paper on an optimistic note, writing that it might still be possible to avoid some of the worst effects. "If we can identify in advance the most vulnerable forests," they wrote, "in some cases management . . . might be able to ease the transition to new and better-adapted forest states." But as Stephenson later told me, that's a big if. A tipping point is where the straight line of events takes a sudden turn. It's a hard thing to see coming. "Oh my gosh, thresholds are sneaky," he said. "Ninety-nine percent of the time we humans are just going to be caught off guard."

Temperatures rise, and there are hotter droughts, more insect outbreaks, and more fires. Forests do not recover from a first blow by the time the second arrives. Once you start to think this way, it can be easier to imagine how forests might be destroyed than to imagine how they might continue. But this is too simple a model, extended too far. There will be trees and forests, probably for a long time. The question is, which trees will be where?

From Saguenay, where Hubert Morin showed me a spruce budworm outbreak, I continued west, past Lake Saint-Jean to the town of Normandin. There I visited a *pépinière,* or tree nursery. My guide was a technician named Frédéric Fortin. He took me first to a seeding facility, where a series of machines filled sprouting trays with dirt, injected each of the tray's chambers with a pair of seeds, then covered the dirt with small white pebbles. We went next to a greenhouse—a white plastic arch as long as a football field—to see that season's seedlings. Each was just an inch tall, a little green starburst. After a year in the greenhouse, the seedlings spend two years outdoors. Fortin took me to see black spruce and jack pine seedlings that were a foot tall, nearly ready to be planted.

We drove out of the *pépinière,* past fields of *bleuets* (still semi-wild, Fortin told me—"You can't plant blueberries. It's impossible."), and across the street to the seed orchard, a field of evenly spaced rows of small trees. This is where the *pépinière* gets its seeds. Fortin told me the *pépinière*'s foresters select the fastest-growing, straightest trees to sire its seedlings, the same way the radiata farmers I met in New Zealand have worked to improve their stock. Pépinière de Normandin produces some ten million seedlings each year. Seedling nurseries are scattered around the province, each supplying local reforestation projects. It is important that each region has a local supply of seedlings. While the seedlings that each nursery produces may look interchangeable, they are—in theory, at least—custom-built for their place.

In the 1750s, Carl Linnaeus noted that yews grown in Sweden from local seeds fared better than yews grown from French seeds. By the late 1800s, foresters throughout Europe were planting so-called common gardens, growing seeds from across the range of a single species together to see which did best. Local seeds usually outperformed seeds from farther away. "In any stand of trees, there has to be intense selection throughout a single generation," said Jerry Rehfeldt, a retired U.S. Forest Service geneticist. "Competition is severe." In a burned-over acre of lodgepole pine, tens or even hundreds of thousands of seedlings might sprout, he told me. Only a thousand or so would survive to maturity. The trees that survived would tend to be those best suited to local conditions. The genes that contribute to success would spread through the local population. For more than a century, the forester's rule of thumb has been to plant seeds from local sources.

But, as foresters Tom Ledig and Jay Kitzmiller wrote in 1992, "If global warming materializes as projected, natural or artificial regeneration of forests with local seed sources will become increasingly difficult." In effect, a common garden is a climate change experiment. Planting a Swedish yew in France simulates planting it in a warmer world. Modern climate change will perform the same experiment on a grand scale. Higher temperatures might mean that a French yew would grow better in Sweden than a Swedish yew would. They might mean that a yew wouldn't grow in France at all.

In a general way, warmer temperatures will push species poleward and upslope. In the short term, foresters could run backward down this escalator: Instead of using local seeds to replant after harvest, they could acquire seeds of the same species from a little way downslope or toward the equator. The provincial governments in British Columbia, Alberta, and Quebec have begun shifting their "seed transfer zones," which dictate where the provinces' foresters acquire and plant seeds in reforestation projects.

But foresters might eventually have to reconcile themselves to planting entirely different sets of species, Ledig and Kitzmiller wrote. Climate change will push species poleward and upslope, but it will not do so uniformly. A changing climate will expand the range of possibilities for some species and shrink them for others. The difficulty lies in predicting which trees will succeed and fail in which places.

The fastest method of prediction is to use models. In the early 2000s, Jerry Rehfeldt and his Forest Service colleague Nick Crookston set out to model the current and future ranges of dozens of tree species in western North America. The Forest Service maintains hundreds of thousands of Forest Inventory and Analysis plots. They are similar in form and function to the plots scattered across Quebec, although the American plots are less regularly spaced. For each tree species, Rehfeldt and Crookston selected plots that contained the species and plots that didn't. They overlaid these plots with a climate model constructed from scattered weather stations in order to compare conditions in places with the species and without it. They used the resulting information to build a model of the conditions under which the species might live. To test the model, Rehfeldt and Crookston compared its projection of the organism's range with that organism's real-life range. An accurate model would predict almost exactly where the species actually lived.

Once the model appeared sure-footed in the present day, Rehfeldt and Crookston sent it into the future, using the Intergovernmental Panel on Climate Change's climatic projections for 2030, 2060, and 2090. The resulting range maps suggest what foresters and conservationists may face in coming decades. As the models move further into the future, some widespread species—like Douglas-fir, quaking aspen, and lodgepole pine—retreat north or upslope. Many others, especially rare species—like Alaska yellow cedar, alligator juniper, and Colorado blue spruce—just fade and disappear.

But some of the models raised another possibility. Brewer spruce, *Picea breweriana,* is a rare species endemic to the Klamath Mountains of northwestern California. The model projected that by 2090, the species would disappear from its current range. But it also predicted that newly suitable habitat would appear in the Nass-Skeena region of British Columbia and in the Yakutat–St. Elias region of southeastern Alaska—hundreds of miles north of its current range. "It is extremely unlikely that the species could disperse and establish that far north in less than a century," wrote Jerry Rehfeldt, Tom Ledig, and Barry Jaquish in 2012.

As Rehfeldt pointed out, there are difficulties with these kinds of models, particularly for rare species like the Brewer spruce. They incorporate only one of the three broad reasons why a species might live where it does and not where it does not. The model is built using environmental conditions, whereas the species' actual range might also reflect its interactions with other species and its luck in reaching otherwise suitable places. This means that such models likely often underestimate the possible future range of a species and may overestimate how quickly a species will disappear from its current range. Range models can't predict all the places where a species might or might not be able to survive, Rehfeldt told me, but they provide a place to start, a hypothesis to test.

The ultimate measure of which trees might grow in which places is still a common garden, where real ground is planted with real trees. In their paper on Brewer spruce, Rehfeldt, Ledig, and Jaquish suggested that people should help the spruce move north to suitable habitat. The British Columbia Ministry of Forests, Lands, Natural Resource Operations and Rural Development would soon establish two common gardens, they wrote, with seeds collected from ten Brewer spruce populations. These populations spanned the entire north–south range of the species, so that "it is likely that most of the species' genetic diversity will be captured." In 2019, I asked Barry Jaquish whether the ministry had followed through on the plan. He told me that it was still pending. "We haven't had the resources to establish the plantings," he said, via e-mail. "It's still in the pipe."

In the early 2010s, a group of foresters in Quebec planted a series of common gardens with white spruces grown from seeds collected across the province. They found that, already, the trees that performed best were those grown from seed collected a bit to the south. In a paper describing their efforts, the researchers distinguished between the various ways foresters might move trees. The method the group suggested for Quebec's white spruce was what they called "assisted population migration," in which seeds from a southern population of trees were planted farther north but still within the species' current range. This was in contrast to "assisted range expansion," where seeds would be planted outside the species' current range. Forest geneticist Jean Beaulieu was a coauthor on the paper. When I spoke with him and his colleague Sylvie Carles on the phone, I asked them about this distinction. Imagining that climate change will continue well into the future and that foresters may face the same problems for centuries to come, I said, won't assisted population migration eventually lead to assisted range expansion? Won't we just have to keep moving the trees, over and over?

Beaulieu stopped me, not unkindly. I was getting ahead of myself, trying to peer too far into the future. The researchers' goals were more modest: to maintain production—forest in, forest out. "We're not trying to save the species, you know," Beaulieu said. "We're trying to save the forest industry."

From Pépinière de Normandin, I drove northeast past Lake Saint-Jean to the Dolbeau paper mill. I'd hoped to see how Quebec's boreal forest is digested into paper. But the communications director I'd been talking with was unexpectedly called to meetings, unreachable. When I approached the people at the front desk, they deliberated, then announced that there was no one who had the time to guide me through the factory.

Unsure what to do next, I drove across the Mistassini River and parked by the road. I sat on the bluffs and looked across the river at the plant. It was huge, the length of a city block: corrugated steel and concrete, with storage tanks, big bay doors, power lines, a water tower, and a smokestack. Up close, the mill had smelled of burnt toast. Steam rose from a dozen vents and, every few minutes, erupted in a geyser from one of the pipes on the roof. The sound would arrive a half second later, a noise like paper tearing. Just to my left, on my side of the river, was the plant's sister, a sawmill. In the yard lay thousands of logs, a whole forest in repose. Most would get sawn into framing studs. The offcuts and sawdust would then cross the river to the paper mill, where they would pass first through a series of chemical stomachs that strip the useful cellulose fibers of their lignin and hemicellulose, then through the forms, presses, and rollers that shape them into paper. The Dolbeau plant produces specialty papers, for magazines, catalogs, instruction manuals, and books.

Different people and organizations use different definitions of "deforestation" and "forest degradation." Foresters and forestry organizations tend not to view timber harvests as deforestation, so long as there are plans to either replant or let the forest regenerate on its own. Ecologists and environmental organizations point out that a clear-cut—however much foresters expect it to become a forest again—neither looks nor behaves like a forest.

The rate of outright deforestation in Quebec, at least, has fallen. In 1992, geographer Arthur Mather argued that most developed countries had experienced what he called a "forest transition," where rates of deforestation had slowed and eventually reversed. In the United States, the forests reached their lowest extent in the early 1900s, right around when William Hall and others warned of a looming timber famine. Mather noted numerous other examples: Forests were expanding in countries across Europe, in Algeria and New Zealand, in Chile, Cuba, Morocco, and China. "This transition appears to be related to a slowing of population growth rates, and to changing attitudes and perceptions on the parts of both peoples and governments," he wrote.

But as a whole, the world is far from making this kind of transition. In many regions deforestation continues at a ravenous pace. In 2015, a group of scientists led by Thomas Crowther estimated that since the early Holocene, humans have felled roughly half the world's trees. Deforestation has shifted to the world's equatorial regions, where the concentration of both carbon and species is highest. As always, much of the deforestation is not for wood but because forests are spaces that can be used for other things—palm oil and soybeans and wheat and maize.

I watched the paper mill for a long time, debating what to do next. Then I got into my car and drove southeast, past the town of Saguenay and on to the St. Lawrence River. Quebec City, where I'd come from, was to the right, but instead I turned left, driving northeast along the water.

As the Sun set behind me, I passed through a string of little towns. Log trucks rumbled south. At the town of Baie-Comeau I turned left again, following the Manicouagan River north. When it was dark, I pulled off the road and slept.

In the morning, I drove on. I passed through a balsam fir forest beset by spruce budworm, trees grayed and cobwebbed for fifty miles. At the mouth of the river I had passed a dam called Manic 1, and soon I encountered more—Manic 2, 3, and 4. The loose weave of firs gave way to scraggled black spruce. Finally I came to Manic 5, now called the Daniel-Johnson Dam, a seven-hundred-foot-tall wall of cement that stretched nearly a mile across a wide valley. Its face was a series of arches that flared into buttresses like cypress roots. Past the dam, the pavement gave way to dirt. I saw all the beasts of the boreal: ravens, magpies, chickadees, spruce grouse, moose, bears, and a porcupine, carefree, eating some ditch plant, quills dragging in the dust.

As I drove north, the land crumpled, folding up on itself. Sometimes past a break in the trees I could glimpse the Manicouagan, now bunched up behind the dam. At midday, I reached Relais-Gabriel, more of a glorified gas station than a town. In a diner, I found two women who told me to drive just a little farther, so I did, up the road to a snowmobiling lodge. Inside, I met David ("da-VEED"), who wore sweatpants and a flat-brimmed hat over his mullet and seemed as if he'd been expecting a reporter from California to turn up at any moment.

David gave me a tour of the lodge, then showed me a topographical map tacked to the wall, a bird's-eye view of the lake. It was a huge circle, sixty miles across, the center filled with an island. Almost two years earlier, as I'd begun thinking about a trip to Quebec, I had spotted this circle on a map. It is called the "Eye of Quebec." The island in the center, Île René-Levasseur, was formed in 1968, when the reservoir behind Manic 5 began to fill. The circle marks where a meteorite hit Earth some 214 million years ago. Researchers estimate that the meteorite was about five miles across, traveling ten miles per second when it hit. It punched three miles into Earth. David showed me a rock from the center of the island. It looked like a broken chunk of concrete, smaller stones welded together by the impact.

After I said good-bye to David, I drove back out to the main road and pulled off at an overlook. I walked to the edge of the hill, the lake and the island spread out before me. When the meteorite hit, in the late Triassic, the land was covered in ferns, ginkgoes, cycads, and conifers, among them the ancestors of sequoias and redwoods. Now it was black spruce as far as the eye could see, stretching away to Alaska. It will change again, but how fast, and to what?

As I'd driven north, hours and hours through the endless forest, I'd wondered a little about what I was after, why I was so drawn to that circle on the map. But now, looking across the water at Île René-Levasseur, at these hills where the Earth was frozen in the torn-open moment of a rock hitting a pond, it seemed like a simple thing. In a world ruled by inertia, by incremental, often imperceptible changes, here was a place where change came all at once.

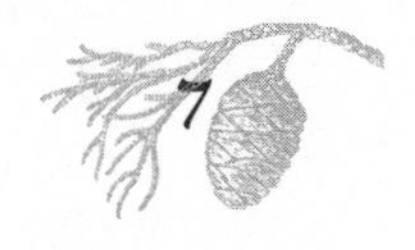

WHERE TO PLANT A TREE

In the fall of 2014, I read an article in the *New York Times* by journalist Jim Robbins. It was about the fate of California's giant sequoias. Three years into the drought, the giant sequoias in Sequoia National Park were losing their foliage. Nobody at the park had seen it happen before.

This interested me. I'd visited the park twice: once just a couple years prior, and once a long time before that. There is a picture in my parents' hallway of my brother and me standing at the foot of one of the sequoias. We were very small. The tree had a catface scar, a big fire hollow at its base that angled closed.

Now, it seemed, the changeless trees were in trouble. Robbins had interviewed Nate Stephenson, the U.S. Geological Survey ecologist. "If there's long-term drought, within 25 years we could see seedlings in trouble," Stephenson told him. "In 50 years, the whole population could be in trouble." In a century, he said, "most of the big trees could be gone."

But what struck me most about the article came about two-thirds of the way down. Robbins wrote that a timber company called Sierra Pacific Industries had begun a "novel program" to protect the trees. It was collecting sequoia seeds and planting them across its holdings, sometimes far north of the trees' current range. "The goal is to conserve the genetic diversity of the native groves," Glenn Lunak, the program's coordinator, told Robbins. And that was it. Robbins moved on to something else.

I was full of questions. Stephenson said that the trees might not be capable of surviving where they were. Was that what the timber company was doing, then—looking for the right place for sequoias? What did the company want with them? Where was it taking them? It struck me as strange and fantastic and possibly important.

In the years since, I've heard many stories like this one—stories of people moving trees, in ways that look by turns noble and mundane and romantic and grandiose. These people have called what they're doing by many different names: horticulture, forestry, *ex situ* conservation, assisted migration. Sometimes these actions seem likely to help either the tree or the people, but sometimes—as ever, when we use human hands to shape that which is not human—they seem just as likely to make things worse.

As the reader may have noticed, this book is as much about people as it is about trees. A tree, after all, is an unparticular thing, just a tall, woody shape of plant. I could have told the biogeographical story of a mismatched world with countless other life-forms—with a quillwort or a pika or a coral as the star, for instance—but such a tale would have stumbled over endless hurdles of understanding. Trees are more familiar. They are everywhere, covering a third of the land on Earth. More than any other living thing, they define their place. Trees can be a texture, like a field of grass, or they can be taken one by one, as individuals with humanlike trunks and limbs. People like trees. Everywhere I go, people take me to see their favorite tree. They stand next to it and give it a slap on the bark and water it with praise. "It's had the shit beat out of it," they say, "but it just keeps chugging along."

Other writers have spent entire books sounding the depths of our symbolic relationship with trees. What most arouses the inner philosopher is not trees' size or shape or ubiquity but their age. There's a good chance any given tree will outlast any given person, so trees form a bridge beyond the usual human experience—a tangible, living link to the past and maybe to the future. The inner philosopher stands in awe. The inner philosopher pens aphorisms and inspirational quotes.

Alexander Smith: "A man does not plant a tree for himself, he plants it for posterity."

John Boyle: "Trees are the best monuments that a man can erect to his own memory. They speak his praises without flattery, and they are blessings to children yet unborn."

Susan Fenimore Cooper: "There is . . . something in the care of trees which rises above the common labors of husbandry, and speaks of a generous mind."

John Evelyn (quoting Virgil): "For what more august, more charming and useful, than the culture and preservation of such goodly plantations: That shade to our grand-children give."

Trees are a powerful way of shaping the world to fit. They provide raw materials, fuel, fruit, shade, and habitat for other species. They slow erosion, dull the wind, produce oxygen, and store carbon. They are often beautiful. For millennia, people have planted trees as a way of helping other people. But it is only recently that we began to imagine that in helping ourselves, we might also help the trees.

Jim Robbins's article on sequoias in the *New York Times* came out in August 2014, but I didn't read it until October. By the time I called Nate Stephenson, he told me it was likely too late for me to visit—it could snow any day, and then the field season would be over. I went anyway. I drove from San Diego, where I lived at the time, and slept in my car beside the road in Three Rivers, just a few minutes from the park entrance. When I called Stephenson in the morning, he told me I was in luck—it had snowed, but not too much. He would be headed up the mountain with the field crew for the final trip of the season, leaving in an hour.

We spent the day hiking through the park's Giant Forest grove. I followed as the field crew measured what portion of each sequoia's crown had died: 0–10 percent; 10–25 percent; 25–50 percent; 50–75 percent; 75–100 percent. We ate lunch on a patch of bare stone overlooking Moro Rock, a huge granitic dome that juts from the mountainside, facing west. I hadn't had time to pack anything that morning, so everyone shared a bit of what they'd brought. Nate Stephenson let me have some of his peanut butter sandwich.

We arrived back at the USGS field station late in the afternoon. I drove through the night to Chico, in the north-central part of the state. In the morning, I met Glenn Lunak at a Sierra Pacific Industries outpost. He had a neatly trimmed gray mustache and wore a flannel shirt tucked into his forest-green pants. He spoke with a slight midwestern accent. He seemed to me like the holotype of a forester.

Lunak and I drove into the hills, joined by two other Sierra Pacific foresters. After spending most of the morning on the dirt roads that wound through the company's forests, we parked on a hillside. We leaned against the truck while we ate our lunch. (Lunak brought an extra sandwich for me.) From up the road came the sound of an engine, someone on a four-wheeler. One of the foresters stepped around the truck and held out his hand. The driver stopped. When he pulled off his helmet, the foresters recognized him as a company biologist; he'd been out in the woods trying to trap fishers, he said, but all he'd gotten were gray foxes.

"So, there's giant sequoias here," he said. "Did you guys plant them?" The foresters laughed and told him yes.

"Huh, pretty cool," he said. He put his helmet back on and drove away.

On the uphill side of the road was a natural forest, thinned but never fully cleared. Below was a clear-cut that Sierra Pacific had recently replanted. The cut was stubbled with ponderosa pines, incense cedars, Douglas-firs, plus species that Sierra Pacific didn't want but grew anyway, like bull thistles and manzanitas and California nutmegs, *Torreya californica*, sister species of the Florida torreya. Scattered among them were several hundred sequoias. Lunak walked down the slope and stopped in front of one of them. Its trunk was lavender, two inches wide at the base. Its blue-green boughs were four feet wide and four high. "Two years' growth and look at that sucker," he said. If all goes according to plan, there will eventually be more than 2,000 of these groves spread across the northern Sierra Nevada and the southern Cascades, covering some 32,000 acres, compared with the roughly 47,000 acres of natural sequoia groves. As the company thins and harvests other trees, Lunak said, it will spare the sequoias, leaving them to grow fat and old.

The company calls it the "Giant Sequoia Genetic Conservation Program." Sierra Pacific started it in the early 2010s, after company vice president Dan Tomascheski attended a meeting of the federal, tribal, and state managers of California's giant sequoias. Tomascheski was invited because Sierra Pacific Industries owns land adjacent to Calaveras Big Trees State Park (where Gus Dowd stumbled upon a sequoia in the winter of 1853) and had helped the park managers with a fuel-reduction project. Many of the grove managers Tomascheski met at the meeting were worried about what a changing climate might mean for the trees. They were discussing the possibility of creating a seed bank, in case the old sequoias began dying off. Tomascheski suggested going one better: Instead of just a collection of seeds, why not a living seed bank, an entire duplicate forest? The company had the space. It was already in the business of planting trees. And it was privately owned, which meant that Tomascheski didn't have to convince shareholders of the scheme, only his boss.

Lunak put together the plan: Over the next two decades, Sierra Pacific Industries would collect cones from each of the seventy-some sequoia groves. The groves are isolated from one another, genetically distinct, so a central goal of the program would be to capture and preserve this genetic diversity. The company would collect cones from trees scattered throughout each grove, to ensure that it was capturing the grove's full genetic breadth. It would keep seedlings grown from the seeds of each grove apart from those of the others, to maintain that uniqueness. Once the seedlings were big enough, the company would plant between three and five new groves in each of its ten districts—or as many as fifty new groves for each original grove. Each of these new groves would be a genetic duplicate of its parent grove. They would be scattered across Sierra Pacific's holdings, the sequoias planted alongside the usual pines, firs, and incense cedars. Some of the groves would be hundreds of miles north of the original groves.

I would come to see it as a textbook example of assisted migration, as carefully planned and orchestrated as any I am aware of. But I wondered why. What was in it for the company? Lunak and Tomascheski insisted the program was altruistic. Tomascheski told me in the fall of 2019 that the program had so far cost the company something like $300,000. The company reveals little about its finances, but it seems likely that the price of the sequoia program is barely more than a rounding error. In 2018, *Forbes* called Sierra Pacific's owner, Red Emmerson, America's "last great lumber baron," estimating his worth at $4 billion, and the company's annual profit to be in the hundreds of millions.

As part of its agreement to collect cones on federal- and state-managed groves, the company has pledged not to sell the sequoias it grows from those groves as timber, but there are indirect ways it could recoup its investment. Once they're big enough, the groves could be entered into California's carbon credit market. Someday the groves could perhaps even be sold to the public as conservation easements or parks. Bill Libby, a retired U.C. Berkeley forest geneticist who helped Glenn Lunak put together the company's sequoia plan, told me he thinks that under the right conditions the sequoia could make a fine timber tree. The trees get brittle when they're big, but when they're young they have mechanical properties similar to those of coast redwood—currently one of the most valuable timber trees in North America. Perhaps the company's new groves are its future seed orchards. Or maybe—by the most cynical read—Sierra Pacific's goal is merely to entice reporters like me. More than one environmentalist I spoke with about the company's sequoia program called it greenwashing, a public relations move.

But it is telling that the people who manage the wild groves have widely agreed to let Sierra Pacific collect cones. The Sierra Pacific employees I met, meanwhile, seemed sincere in their intentions. In the morning, before our visit to the new sequoia grove, Glenn Lunak took me to see some sequoias he'd planted back in 1980, when he was the area's replanting manager. "I think I bought five hundred giant sequoia seedlings, planted them just to see how well they would do," he said. "It's been years since I've driven back here." We pulled into a clearing. Lunak hopped out of the truck and looked around, taking in decades of change. This plot had been clear-cut and replanted, but now the woods looked almost natural, a mix of pines and firs and incense cedars, plus the sequoias. The sequoias were seventy or eighty feet tall, many of them with trunks more than two feet wide. "This is way cool!" he exclaimed, and laughed.

He wandered into the woods, dead branches snapping under his feet. He stopped at a sequoia growing on a slope. The tree's bark was lighter than that of the old sequoias, the corrugations more regular. Its canopy tapered in a perfect Christmas-tree cone, a spearhead reaching skyward. "It's only thirty-four years old," he said, looking up at it. "I guess they're growing pretty good, by God!"

As we drove in, Lunak had quoted Wesley Henderson. In the book *Under Whose Shade,* Henderson wrote that on his high school graduation day, his father, Nelson, a second-generation farmer in Manitoba's Swan River valley, said to him, "The true meaning of life, Wesley, is to plant trees under whose shade you do not expect to sit." Lunak paused, then added, "I take that to heart." As we drove back out of the grove, he was quiet. The woods on either side were tall. Sunlight flickered through the shade.

I wrote an article about the Sierra Pacific Industries project for *Guernica* magazine. But even after the article came out, I kept thinking about trees being in wrong places, in less than ideal places, and maybe sometimes in perfect, paradisiacal places. Before long, I found myself on the phone with Connie Barlow.

In the winter, three years after our first conversation, I met Barlow and her husband, Michael Dowd, in the northeastern corner of Alabama, at the edge of the Appalachians. They were roosted there for the season at the cabin of a friend. I arrived after dark. They both came outside and greeted me with hugs. Barlow had short, gray-brown hair and wore sweatpants and a vest. She explained that the lisp I heard in her voice was because she'd just had three teeth removed. She grinned to reveal the space in the lower half of her smile. "I look like a vampire bag lady," she said. Dowd was tall, with a deep voice and big brown eyes and a beard that traveled from ear to ear without crossing his upper lip. In the time I spent with them, I learned a few of their habits. They called the moon "Luna," and the large beech by the creek "Beechy." They called their van "Angel," and their relationship "Jasmine." That way, they explained, when they had arguments they could ask what was best for Jasmine rather than what was best for Connie or Michael. When they had small squabbles over whether to leave the heater on or off, Jasmine went unmentioned.

During the day, Barlow and I wandered through the forest. The property was bisected by a creek. There were uplands and floodplains, clothed in a mix of pitch pines and slash pines and a wide array of hardwoods—oaks, beeches, sweet gums, hickories, umbrella magnolias, and liriodendrons, which, Barlow reluctantly explained, are commonly called yellow poplars or tulip poplars, this despite the species being neither a tulip nor a poplar.

The trees were big. Later, the owner of the property told me that the forest hadn't been logged for at least a century. She'd moved there from Boca Raton in 2008, after her husband died. He loved sailing, she said, but she liked the woods. The property had been cheaper than the surrounding lots, which the real estate agent told her was because it contains a conservation easement, managed by the Nature Conservancy. The conservancy wanted to protect the property because its forest was intact and because it held a rare species, the sun-facing coneflower. The agent thought the restrictions on the easement were onerous, but the owner told me she thought it was perfect. "Don't throw me into that briar patch!" she recalled thinking, and laughed. There were thirteen pages of restrictions. Among them: no logging, no subdividing, and no planting nonnative species.

For hours, Barlow and I splashed back and forth through the creek. We looked at trees and considered their arrangement. We deliberated for a long time about whether or not a beaver was responsible for a chunk missing from the base of one tree. She thought so; I thought not. Barlow places much weight on firsthand observation. People had a broader view of things, she said, before naturalists turned into scientists. Sometimes she filmed what we saw with a handheld video recorder, narrating the scene. Although she used to write books, she is now convinced that YouTube is the true library of the people.

We spent much of the day pulling privet, *Ligustrum sinense*. It grows as a bush, sometimes a tree. In its native China, it makes perfectly orderly hedges. In the American South, it makes all kinds of shapes, none of them pleasing to Barlow. It has leathery leaves and clusters of black berries. Beyond that pale description, I'm unsure. It turns out I suffer from near-total privet-blindness. As we walked, Barlow stooped down repeatedly to pluck the small privets I'd missed. She hung them upside down from the branches of other trees, so that their roots would dry out. We arrived finally at a corner of the property where I couldn't help but notice the privets. They were so thick they had almost reverted to hedge form. We attacked them with a Pullerbear, a shoulder-high length of square steel tubing attached to a set of jaws. The jaws go around the base of the privet's trunk and clamp down when you weight the handle. You lean on the handle and yank the privet from the ground, roots and all. Sometimes the privets were tougher and we had to use a saw. She doused the sawn stumps with herbicide.

The potential irony of her anti-privet campaign was not lost on Barlow. After spending the last decade moving a tree to places it hadn't managed to reach on its own, she was now hard at work killing trees that someone else had moved. But she saw them as different cases: It was one thing to move the Florida torreya to places where it plausibly might have lived in the recent geological past, she said, and another to move a privet to an entirely new continent. A note on the Torreya Guardians website urges people not to plant Florida torreya west of the Mississippi.

We spent hours Pullerbearing and sawing and hanging privets from trees. "Some people like to meditate to empty their mind," Barlow told me. "I think priveting is best." Sweating from the effort, I stripped to my shirtsleeves. There were dead privets everywhere. There were live privets everywhere. "I love the futility of it," she said. There was also scattered trash, but she ignored it. "I don't pick up the litter," she said, "because the litter doesn't reproduce."

In the late afternoon, Barlow, Dowd, and I went for another walk in the woods. We passed Beechy the beech and a little pond, and crossed the creek a couple of times. Dowd led us over a small rise. There, poking from the dead leaves, was a tiny evergreen seedling. It was a Florida torreya.

In the years since she'd founded Torreya Guardians, Connie Barlow's views had taken an apocalyptic turn. When she first began to think about moving the Florida torreya north, she was interested in climate change but viewed it as largely separate from Paul Martin's idea of rewilding, which was about re-creating ancient ecological partnerships undone by humanity. Then, one morning in December 2012, Barlow and Michael Dowd experienced what they call their "climate awakening." It happened after they read an article in the *Atlantic* titled "5 Charts About Climate

Change That Should Have You Very, Very Worried." The charts were about rising seas; dying coral reefs; increasing wildfire, drought, and war. Barlow and Dowd got so worried that they wept. Barlow struggled for a long time afterward to accept the implications of climate change. Denial, anger, bargaining, depression. Finally, she accepted the world's fate. She accepted the end of civilization. "I'm a collapsitarian," she told me cheerfully. "I really think things are going down."

During this period of realization, she said, planting trees provided her with emotional ballast. But her expectation of the imminent collapse of society also heightened her alarm over the fate of the world's forests. It meant that time was running out. Ideally, she said, thousands of people could each pick a species, learn as much as they could about it, then try moving it to where they thought it might need to go. These people should do everything they could to avoid introducing new invaders, she said; "You don't bring in species from another continent, you just don't do that." But a fear of making mistakes shouldn't stop people from experimenting, she said. It was too late for that.

Torreya Guardians provided the template. By the time I visited Barlow, the organization she'd started was fifteen years old. It had sent Florida torreya seeds to Georgia, Alabama, South Carolina, North Carolina, Virginia, the District of Columbia, Kentucky, Tennessee, Ohio, Michigan, Illinois, Wisconsin, Massachusetts, Pennsylvania, New Hampshire, and Vermont. Some of these seeds had grown up to produce seeds of their own. Another member of the group, Paul Camire, had compiled a list of dozens of Florida torreyas that people had planted in the 185 years since Hardy B. Croom encountered the species. Camire tallied Florida torreyas in arboretums and botanical gardens, in public parks, and on private property, as far away as Seattle, Washington, and Edinburgh, Scotland. Many of these trees had also produced seeds. In several places these seeds had grown into seedlings. The torreyas growing wild in Florida's Torreya State Park were unlikely to recover, Barlow thought, but the species as a whole was no longer in immediate danger.

In the summer of 2019, she sent the U.S. Fish and Wildlife Service a petition to downlist the Florida torreya under the Endangered Species Act, moving it from "endangered" to "threatened." Downlisting would declare the Torreya Guardians' efforts a success and would encourage other people and organizations to launch their own efforts to save rare plants, she wrote. The government didn't need to spend decades and millions of dollars doing something that volunteers could do perfectly well on their own. "Federal taxpayer money," she wrote, "could be freed up for recovering endangered species that aren't as easy to work with as putting seeds in a pocket and digging little holes in the forest."

Jason Smith, the University of Florida, Gainesville, plant pathologist, called Barlow's effort to downlist the tree absurd. "We're talking about North America's most endangered tree species," he said. If she succeeded, he said, it could make things harder for everyone else working to save the tree. Barlow had become obsessed with one conservation tool at the expense of all the others, he said. "I don't think she cares about torreya. I think she cares about assisted migration." (As of November 2019, Barlow's petition was still under review, according to the response letter the service sent her.)

It has now been more than three decades since scientists first suggested that many species would not be able to keep up with the rate of modern climate change and that humans might help them move poleward or upslope. Assisted migration—or "assisted colonization" or "managed translocation"—remains the subject of much debate. More than a hundred papers on the subject have been published in scientific journals. Some of these deal with specific life-forms—New Zealand's tuatara, Iran's Farsi tooth-carp, the American pika, and many, many trees—while others treat the subject more generally. The question, still, is whether the risk of inaction outweighs the risk of action.

Many scientists remain worried that assisted migration will do too much—that it will create new invaders, alter or damage recipient ecosystems, or confuse a public finally beginning to absorb the risks of rearranging the world's living things. Some are concerned about who might take such an action. "This is not gardening," ecologist Thomas Abeli told me. "I think assisted colonization should be strictly supervised by ecologists and conservationists." Dan Simberloff, the invasion biologist, told me that assisted migration will also do too little. Set against the vast changes that humans have wrought upon Earth, assisted migration "is a Band-Aid at best," he said. "It gives us a false sense that we're doing things that matter."

So far, there are few examples of people actually moving trees—or other species—poleward or upslope as a conservation method. I am aware of only a couple dozen such projects—a butterfly here, a common garden there. Even fewer are explicitly billed as assisted migration. There is nothing like the popular movement Barlow envisioned. But Mark Schwartz told me he thinks that eventually more people will be tempted. "People want to do stuff," he said. "They want to take positive action." He told me that his thinking on assisted migration had evolved in the years since he wrote the original argument against Connie Barlow and Paul Martin's proposal to move the torreya north. "I've become more sympathetic to this idea, this problem," he said. "The climate is going to continue to change." But he said he still worried about what might happen if people began widely following Barlow's example. It was possible, certainly, that a single person could save a species. But it was also possible for that single person to do great harm. "Look," Schwartz said, "I understand that we're likely to be doing this more and more in the future. I think we should be doing it thoughtfully."

Schwartz is part of a team of scientists currently working on a set of risk-assessment guidelines for National Park Service employees to use when considering assisted migration projects. "If you want to move a bull trout or blue butterfly or giant sequoia, what specifically would you do?" he said. In September 2019, he sent me the latest draft. The eighty-page document asked users to rank the risk, from "low" through "very high," of a long list of possible actions: What is the risk of action or inaction to the species in question, and to the source ecosystem and the recipient ecosystem? In separating one population of a species from another, would the species' would-be saviors spark undesired evolution? Does the species have the traits of a likely invader? Finally, is the species of great value to society? Reading through the guidelines, I'd expected them to end in a calculation, where all the judgments of low risk and medium risk and very high risk would be added up to some score, some definitive way of deciding whether or not it was worth moving a species to try to save it. But there was none.

Now, in the northeastern corner of Alabama, Connie Barlow, Michael Dowd, and I wandered through a grove of tiny torreyas. Barlow commented on their condition, on the slope and aspect and soil conditions where they were growing, and on the effects of browsing animals. "Oh God! It was beautiful!" she moaned at the sight of one, once six inches high, now trimmed to a nub by some herbivore. She reconsidered. "Well, okay, this is good," she said. "We can learn from this."

She was a little chagrined. She hadn't meant for me to see these seedlings. Dowd had led us to them without consulting her. Her friend—their host—didn't know Barlow had planted torreyas on her property. It violated the terms of the conservation easement: no nonnative species. Later, I asked Barlow why she had done it. She had already sent torreya seeds all over the eastern United States, to people who had willingly and legally planted them on their land. Why had she chosen to plant them there, risking the easement and probably the friendship?

She answered slowly, as if struggling to translate feelings into words. Soon, she said, the issue would be moot. People would accept the reality of climate change. They would understand what the fossil record shows: that when the climate changes, the arrangement of the world's living things changes with it. They would see the ceaseless journeys of trees and accept the idea of assisted migration as a matter of course. The concept of conservation easements would change to reflect this. The forest around us was beautiful, but it would not always be this way. "Acceptance of assisted migration," she said, "means you accept that eventually your child is going to move away."

The day after my visit with Connie Barlow and Michael Dowd, I drove south to Florida. As I neared Torreya State Park, I began to see broken trees, flattened trailer homes, abandoned cars, and drifts of debris. Two months earlier, Hurricane Michael, a Category 5 storm, had dragged across the Panhandle. The park took a direct hit. I found it in disarray. Before the storm, the limestone bluffs overlooking the Apalachicola River were covered in dense forest. Now the view was open on all sides, the foreground a tangle of downed trees. Scientists estimated that some 750 Florida torreyas remained in the wild before the storm. When I visited, they were still looking for survivors.

In the Talmud there is a story of a man named Honi Hama'gel, who was out walking when he came upon a man planting a carob tree. He asked the man how long it would take for the tree to bear fruit. "Seventy years," the man said.

"Art thou, then, sure that thou wilt live seventy years?" Honi asked.

"I found carob-trees in existence when I came into the world," the man said. "My ancestors must have planted them. Why should I not also plant them for my children?"

Soon after, Honi fell asleep, and in a Van Winklian turn, he continued to sleep for seventy years. When he awoke, he saw a man picking carob fruit from a nearby tree. "Didst thou plant this tree?" he asked the man.

"Nay," the man said. "I am the grandson of the man who planted this tree."

A person plants a tree, shapes a bit of the world to fit, and so extends a gesture of goodwill to the future. It would feel like farmer-gardener propaganda to anybody but a farmer-gardener. Today, people plant trees with a new aim, hoping to make more favorable not only the local environment but the global one. "We know that the air in a close apartment is appreciably affected through the inspiration and expiration of gases by plants growing in it," wrote George Perkins Marsh in *Man and Nature*. "The same operations are performed on a gigantic scale by the forest, and it has even been supposed that the absorption of carbon, by the rank vegetation of earlier geological periods, occasioned a permanent change in the constitution of the terrestrial atmosphere." Now we release this long-stored carbon, clearing and burning the forests of today and digging for the coalified remains of forests past. This carbon traps the Sun's heat, warming the planet. But new trees could replace the old ones, and perhaps limit the damage.

A nine-year-old German boy named Felix Finkbeiner began what is currently the largest carbon-focused tree planting campaign in 2007. His teacher had given Finkbeiner's class the homework assignment of writing a research report on climate change. Finkbeiner told me he was inspired by the story of Wangari Maathai, a Kenyan professor of biology. In 1977, Maathai began a tree-planting campaign called the Green Belt Movement, which paid rural Kenyan women to plant trees near their communities. The campaign had two goals: to slow rampant environmental destruction in the countryside and to provide the women with meaningful work outside the home. In 2004, Maathai was awarded the Nobel Peace Prize for her work. By then the campaign had planted tens of millions of trees.

As a fourth grader, Finkbeiner said, the part about empowering women was lost on him, but he understood planting trees. He stood in front of his class and told them that, while adults dithered in the face of climate change, children could do something about it. Kids, he said, ought to plant one million trees in every country on Earth. "That was the biggest number I could come up with, or something," he said. Two weeks later, the kids got started. The class gathered outside the school and planted a crab apple.

Two journalists were in attendance. They reported what they'd witnessed to the German public. Before the year was out, children across the country had planted fifty thousand trees. At that point, Finkbeiner said, the kids decided to hold a press conference about the campaign, which they now called Plant-for-the-Planet. During the train ride to the conference, his dad told him not to be disappointed if nobody showed up. Finkbeiner didn't have to worry. The conference spawned hundreds of news reports. By the third year, the movement had planted a million trees.

Wangari Maathai was then leading a new tree-planting campaign, this one under the auspices of the United Nations. It was the Billion Tree Campaign, and as with Plant-for-the-Planet, its focus was on trees as carbon storage. Then, in 2011, Maathai died. Achim Steiner, executive director of the U.N. Environment Programme, asked Plant-for-the-Planet to take over the Billion Tree Campaign. The kids agreed. The campaign has since subsumed more regional campaigns, and its tally now includes the trees that various countries pledged to plant under the 2015 Paris Agreement on climate change. As of this writing, the campaign has planted 13.6 billion trees.

People have suggested other ways of slowing climate change: We could reflect the Sun's rays with giant mirrors placed into orbit. We could use machines to suck carbon dioxide directly from the atmosphere and sequester it underground, or we could fertilize the ocean with iron in the hopes of spurring algae growth. Compared with these solutions, replanting the world's formerly forested places with trees seems both practical and innocuous.

But there are ways this, too, could go wrong. A tree planted in the wrong place could become an invader. A tree planted too many times, too closely together, also might not perform as desired. Recently, a group of researchers examined the tree-planting plans that forty-three countries in the tropics and subtropics had submitted under the Bonn Challenge, which aims to reforest 350 million hectares (865 million acres) by 2030. They found that, of the promised trees, nearly half would be in plantations, most of them of the genus *Eucalyptus* or *Acacia*. These monocultures store far less carbon than natural forests, the researchers wrote, and are likely to reduce biodiversity, not help maintain it. A tree could even become a tool of neocolonialism. While Finkbeiner told me that tree-planting campaigns could provide employment and fight poverty, it's also easy to imagine how such campaigns could be carried out not for the sake of the people who live where the trees are being planted, but in spite of them.

Most of all, there is the problem of the changing climate. Trees not only affect the climate but are affected by it. If people plant the same trees in the same places where they currently exist, they will likely suffer the same effects of drought, insects, and fire that have begun to afflict existing forests, David Breshears told me. "I don't think that tree planting is stupid, or that people shouldn't consider planting," he said. "But there's a pretty big, important cautionary footnote that people have to be aware of." Tree planting must be done carefully, he said, and it can only be part of the solution. "If you're not reducing carbon emissions," he said, "everything else is going to be overwhelmed."

Finkbeiner agreed. "Tree planting is not a solution to climate change," he said. "It's just part of the solution." Finkbeiner is now a PhD student in ecology at University ETH in Zurich, Switzerland, in the lab of British ecologist Thomas Crowther. Other researchers at the lab have estimated that Earth could hold some 0.9 billion hectares (2.2 billion acres) of additional canopy cover, which translates to about 1.2 trillion trees. Plant-for-the-Planet's current goal is one trillion trees. But even this would not make up for all the trees that have been lost. "Even if we restored all of those one trillion trees, it could only capture about a quarter of human carbon emissions," Finkbeiner said. "We need to do much more than tree planting."

Many of the people I've spoken with worry that we won't be able to avoid disaster. They have begun to think that things will not turn out all right, not even in our lifetimes, not even for those of us insulated by wealth and geographic privilege. The seas will rise and crops will fail and forests will burn. Collapsitarianism has crept from the fringes. Evolutionary biologist Daniel Brooks provided me with one of the most vivid of these nightmare scenarios. He studies emerging diseases—that is, parasitic organisms that have shifted from one host to another. As climate change allows living things to survive in new places and alters the relationships between species, he said, the odds increase that humans will encounter a new, deadly disease. "The time is short, the danger is great, and we are largely unprepared," he told me. He thinks it's likely that some new disease will kill off much of the world's human population by the middle of this century.

There is also, still, the specter of nuclear warfare. The possibility that we might blow ourselves to bits in some fit of machismo is the impetus behind another tree-moving scheme I recently heard. Its author is Bill Libby, the retired U.C. Berkeley forest geneticist. I first interviewed him several years ago, when I began researching Sierra Pacific Industries' giant sequoia project. He'd helped Glenn Lunak plan the project.

Libby has wispy white hair, limpid blue eyes, and a deep, booming voice. Sometimes when I've visited him at his house in Orinda, California, I've found find him wearing a tattered sweatshirt. This way, he told me, when his Senegal parrot perches on his shoulder, he doesn't have to worry if the parrot relieves himself. The parrot is older than I am. Along with pictures of children and grandchildren, Libby's house is decorated with coast redwood and giant sequoia paraphernalia. His yard is filled with arboreal curiosities, among them a redwood grown from a seed that went to the moon aboard *Apollo 14*—"which apparently did it no damage," he said. The tree is more than one hundred feet tall.

One morning in the winter of 2017, I went to breakfast with Libby and a few of his friends at a diner near the retirement community of Rossmoor, in Walnut Creek, California. The group included a retired Chevron executive named Lloyd, his daughter Caroline, and Manfred Lindner, Barney Rubin, and Gus Dorough. The latter three were all in their nineties. Lindner and Dorough worked on the Manhattan Project during World War II, helping to develop the plutonium bomb that the United States dropped on Nagasaki. Rubin later helped develop and test the hydrogen bomb. We had coffee and eggs and hash browns and talked about the end of the world.

Rubin and Dorough had both witnessed the first test of the plutonium bomb. “It was spectacular, something you can’t believe,” Rubin said. “It was dark, early in the morning, and then suddenly it’s daylight.”

“That was a total surprise,” Dorough said. “I didn’t expect it to turn instant daylight.” The mushroom cloud turned a series of colors, then the ground shook, then there was the sound. “It scares the hell out of you,” Dorough said.

It felt incongruous to think of these frail, pleasant old men helping to sow the seeds of such destruction. It made me wonder, too, how people in the future will look back on people alive today, how they will understand our heedless consumption. I asked the old men whether they regretted what they had been a part of.

“I always worried about it,” Rubin said. “I asked myself, ‘Is it proper to be working on these damn things?’ ” But he reasoned that if it was okay to be an infantryman, okay to carry a rifle, then it was okay to help build a bomb. “Once you’ve established the principle,” he said, “you only have to establish the price.”

Lindner’s answer was even less reassuring. In 1939, Albert Einstein wrote a letter to President Franklin Roosevelt, warning him of the potential that uranium could be used to build a weapon. “It is conceivable—though much less certain—that extremely powerful bombs of a new type may thus be constructed,” Einstein wrote. Knowledge that the atom bomb was possible made it an inevitability, Lindner said. “A lot of the thinking was, ‘If making these weapons is possible, then we have to have it.’ ”

Bill Libby thinks that, just as the knowledge of nuclear weapons made their existence inevitable, their existence makes it inevitable that they will eventually again be put to use. The Northern Hemisphere is home to the majority of the world's humans and to most of the nuclear-armed nations. What this means, Libby said, is that to really save something, you may have to send it south.

He explained it like this: When he was a young man, many decades ago, he went for a walk above the U.C. Berkeley campus. He passed first through a stand of pines. It was a hot day, and the pines gave off a dry, resinous smell. He walked on, into a grove of coast redwoods. "The temperature went down and the humidity went up, and all of the sudden, I just felt better," he said. "It could've been simply that it was a nicer environment in there, but it may have been more than that." Trees emit many organic compounds into the air. There is little solid science on what effects these compounds may have on people, but Libby has heard of others who experience what he does when they enter a grove of coast redwoods. "A percentage of them—and I have no idea what that percentage is, maybe as low as ten percent—report this feeling of well-being," he said. "And that's not insignificant. It might be a worthwhile thing to leave."

He wants to create new coast redwood and giant sequoia forests in the Southern Hemisphere—in New Zealand, southeastern Australia, South Africa, Chile, wherever they will grow and people will have them. Each new grove would contain the entire genetic breadth of the species. "Let them mix it up, sort it out," he said, "and the trees that are better adapted are probably the ones whose offspring are going to colonize the area and create their own little races."

During my trip to New Zealand, I saw a preview of what these forests might look like. When Wink Sutton, the radiata pine breeder and economist, picked me up in Rotorua, the first place he took me was the Whakarewarewa Forest. It is a large grove of coast redwoods, planted more than one hundred years ago as part of the same round of forestry trials that eventually crowned radiata pine. The biggest redwoods were already four feet in diameter and two hundred feet tall. They looked almost identical to those in the ancient groves of northern California, except that here the spaces between the redwoods were filled with forty-foot tree ferns. I can't speak to any particular feeling, but something there draws people. Whakarewarewa, an overgrown plantation, is now among the most visited forests in New Zealand. There were signs instructing people not to pet the trees. Like giant sequoias, coast redwoods have soft, fibrous bark, which people evidently found irresistible. The trees along the path had developed distinctive waistlines from people stroking them.

This, or something like this, is what Bill Libby wants to save. It seems to me both audacious and stunningly small. Libby imagines half the world burned, all the works of civilization undone, countless species and ecosystems and landscapes lost. Into this future he wants to send, above all else, a couple species of particularly large Californian trees and, maybe, the feelings they stir. It's pure romance, of dubious conservation or ecological value, just an old man's gift to a future he won't see. I hope he pulls it off.

I went cone collecting in the fall of 2018. Collecting cones is itself an easy task. Grab the cone, put it in your bucket. The hard part is the logistics. The cones tend to be at the tops of trees, some of them very tall trees. The cones up there are fresher, the seeds they hold more likely to sprout. For a forester to rely on the cones that fall to the ground would be like an orchardist relying only on fallen apples. This is why there are cone collectors. It is a seasonal occupation. Most of the year, the cone collectors I met were arborists and rock climbers. Linemen or window washers would be well suited, too. The cone collectors were used to being up high. They called it "exposure." They were paid by the bushel.

The year before, I'd watched as a crew of cone collectors plucked cones from the blue-green boughs of a giant sequoia, two hundred feet above. The collectors were working for Sierra Pacific Industries. I was sitting with Glen Rouse, the forester who took over the company's giant-sequoia assisted migration project after Glenn Lunak retired. We were pulling the peduncles off the cones. These are the little stems that connect the cone to the rest of the tree. When I first heard the word I assumed it was a forester joke, but later, when I looked it up in a biological dictionary, there it was, right between "pedipalps" and "Peking man." Peduncles gum up the seed extraction process. Rouse didn't want them going in the bag.

Pulling peduncles off cones is boring, and I found that I was envious of the men working high above, envious in a way that was hard for me to explain. Until that day, I'd had no particular interest in climbing trees. As a kid, I was subterranean, never arboreal. I liked to dig holes in sand, dirt, snow, to crawl around in tunnels, to lurk in dark spaces. In first grade, when a classmate bloodied his lip kissing a spruce tree, I couldn't understand his attraction. But that day, watching the cone collectors, suddenly, desperately, I wanted to climb a tree.

Later, I called Robert Beauchamp, the leader of this band of cone collectors, and asked him whether he might let me join. After some cajoling, he agreed. Dan Tomascheski at Sierra Pacific Industries did, too. For insurance reasons, I would technically be a subcontractor. The next fall, a cone collector named Mark Leffler taught me the basics at his rock-climbing gym in Davis, California. Later, I practiced with Beauchamp at his woodworking shop in Woodland, twenty minutes northwest of Sacramento. We slung a line over the limb of an oak tree and I climbed up to it a dozen times, making sure I grasped the fundamentals of switching from an upward to a downward direction.

How it worked was this: I wore three ascenders, one-way cams that let rope through when you slide them upward and that lock in place when you pull back down. One was strapped to my left foot, and one went in each hand. All three were attached by short lines to my belt harness. In a motion like a mime walking up invisible stairs, I could then hoist myself up the rope.

On that morning in the fall of 2018, the cone collectors assembled at a trailhead deep in the Sierra Nevada. When I arrived, they were already laying out their gear between the pickup trucks: olive duffels, five-gallon buckets, coils of rope, throw bags, harnesses with carabiners and ascenders, crossbows, and reels and bolts. Once Robert Beauchamp was satisfied that everything was in order, we hiked into the woods. I followed him down a steep hillside and across a stream flowing over slick granite. We crossed just below a log lying across the streambed. It was eight feet in diameter, gray with age. Past the stream we turned back uphill. I spotted a tiny sprig of blue-green, a sequoia seedling, then another, and another. Farther up were the trees themselves, red-orange, thick, furrowed, squashed around their bases like melted-down candles. Far above were blue-green clouds of foliage, laden with cones.

Glen Rouse had already chosen the trees whose cones he wanted. I sat and watched as Beauchamp made repeated attempts to shoot a crossbow bolt over one of the tree's top branches. The bolt was attached to a reel loaded with fifty-pound-test shark line. Over and over the bolt bounced off the tree. Each time, Beauchamp would swear and reel it back in. Finally it went over a branch and dropped down the back side of the tree. He used the shark line to pull a one-eighth-inch braided nylon tagline over the branch. He used the tagline to pull up a twelve-strand climbing rope. Beauchamp tied off the rope, slipped into his climbing harness, and slowly walked his way up into the tree. When he was almost to the top, he called down to me. It was my turn.

I was sweating by the time I made it twenty feet. At thirty feet, I was higher in a tree than I'd ever been. At fifty feet, I entered what climbers call the "kill zone"—the point past which a fall is likely fatal. Before, I wondered whether I'd be scared of heights, but now I found that I was fine unless I stopped to think about it. At a hundred feet, I reached the first branch. When I hugged it, my fingers barely touched. I felt full of love for it, this fellow organic being, both of us surrounded on all sides by the inorganic, the two of us more alike than not. I also felt a little exposed. As I continued upward, the branches grew smaller and closer together. Finally, an hour after I left the ground, I reached the last branch, 215 feet up. I could see all the way around. To the east, the long sheer wall of the

Sierra Nevada stretched into the distance. The open land to the west faded into haze. The ground was hidden in the boughs.

By the time I arrived, Robert Beauchamp had already half-filled a five-gallon bucket. I sat on a branch as thick as my arm, tied myself off to the trunk with a nylon strap, and transferred the climbing rope from my ascenders to a rappelling device called a Grigri, which, when the time came, would allow me to control the speed of my descent. Beauchamp reminded me of the cones Glen Rouse had asked us to collect, the ones Sierra Pacific had found to contain the most viable seed—green, but beginning to go yellow and brown in places, and with brown peduncles. The technique is to grab the cone firmly and give it a little twist as you pull away from the branch. If you do it right, the branch resists, and you get the cone without the peduncle. But most of the time I did it wrong and had to stop and twist the peduncle off, which was annoying because it took both hands, and for no real reason I'd been keeping one hand firmly latched around a branch.

The bucket gave hollow thunks as we tossed in the cones. When it was full, we transferred the cones to an olive-green duffel, taking care to avoid what Beauchamp called the "rain of pain." This is what happens when you miss the bag and have to pick more cones. We filled a second and a third bucket, enough for the first of two duffels. Before we lowered it from the tree, Beauchamp made me bellow "BAG OUT!" to alert the people on the ground. Then, noting the dwindling supply of cones around my perch, he suggested that I scoot farther out on my branch. He'd already suggested this, and I'd deferred, but now there was no avoiding the move. "Spread your climbing wings," he said. I undid the strap that secured me to the trunk, pulled slack through the Grigri, and slid out on the branch. Close to the trunk it was solid. Now it swayed under me.

But here were cones. Good cones, beautiful cones, bunches of cones. I used both hands to grab them. By then I had mastered the technique, and the peduncles stayed behind. We became connoisseurs, cone-isseurs. "There's the money cone, right there," Beauchamp said, holding one up.

It was the size of a golf ball, a Fibonacci spiral of puckered lips, green with brown and yellow splotches. It was perfect. He tossed it into the bucket.

We fell silent. There was only the soft sound of the breeze and the thunk of the cones in the bucket. My thoughts floated free of their moorings. I imagined the feelings of the tree: no sight, no sound, no smell. It knows the trees around it by the touch of their roots. Maybe there is the sensation of fullness, the pull of water through its trunk, the daily sweep of the Sun, the slight push of the wind. Maybe it could feel the tug as we pulled away cones.

In its hundreds or thousands of years, the tree we were sitting in must have produced millions of cones and hundreds of millions of seeds. These seeds drifted down, grabbing hold wherever they could, each a journey made at random. Perhaps, if the tree is lucky, by the end of its long life one or two of its offspring will have grown up to produce cones of their own. In this way, by the success and failure of countless little journeys, the tree's ancestors had migrated here, come down from the north, across the continent, through the mountain passes, here to this patch of the Sierra Nevada.

Later, the cones we were collecting would go to the dryer. The seeds would go to a nursery. From there, they would go north and upslope, on journeys uncoupled from the route hewn by their ancestors, to places unshaded by sequoias for thousands or millions of years. Some, perhaps, would even go to the Southern Hemisphere, to groves where they might survive the apocalypse, in whatever form it takes. Maybe, much later, someone would stumble upon them, giants lost in a sea of lesser trees, and gaze up at them and marvel that the world could hold something so big and ancient and changeless.

ACKNOWLEDGMENTS

I could not have written this book without help from many people. I would like to first thank my sources. I interviewed more than two hundred people during my research, about a third of whom I have directly quoted. I am thankful to all of them for sharing their time and expertise. A few were particularly giving. Nate Stephenson took me along on two trips to Sequoia National Park and spent much more time on the phone and answering my e-mails. Miriam Jones gave me my first (and still the best) lesson in paleoecology and loaned me a pair of hip waders to boot. Connie Millar invited me to join the GLORIA White Mountains crew, and Brian Smithers answered all my questions about *Pinus flexilis.* Wink Sutton spent two days teaching me about radiata. David Roberts retraced his steps in pursuit of the emerald ash borer, and proved an able reporting wingman when I went door-to-door, looking for quotes. Leah Bauer and Therese Poland helped fill in the story of the emerald ash borer's early days and taught me all about very tiny wasps. Louis Duchesne showed me a new way of seeing a tree, and introduced me to colleagues across Quebec. Emily Coffey and Ron Determann at the Atlanta Botanical Garden showed me the *ex situ* torreya collection and a great many other interesting plants. Bill Libby shared far more wisdom than I was able to absorb. Mark Leffler taught me to how climb a tree in theory, and Robert Beauchamp helped me put my skills to the test. Glenn Lunak, Glen Rouse, and Dan Tomascheski at Sierra Pacific Industries were consistently open and patient through nearly five years of questions. Their work provided the germ of this book. I would especially like to thank Connie Barlow, with whom I do not always agree, but whose conviction, curiosity, and love for the world I deeply admire. Thanks also to Michael Dowd. I will be expecting more e-mails.

Many other people helped with the writing. Lauren LeBlanc and her colleagues at *Guernica* edited and published my first article about trees. Eric Simons provided guidance, reassurance, and edits throughout the process of expanding that article into a book. Adam Hochschild and Michael Pollan offered advice about the publishing world and taught me much about the craft. Kara Platoni provided me with the blueprint for my book proposal. Diana Finch, my agent, believed in the book from the beginning, and helped frame it. Quynh Do, my editor, often saw shapes and patterns that I could not, and made the book immeasurably better. She and the rest of the team at Norton have been a pleasure to work with. Copy editor Bonnie Thompson saved me from many large and small mistakes. Bob Weeden, Mateo Hoke, Jody Tinsley, Katia Savchuk, Annie Zak, and Sharam Baradaran all read partial or full drafts. John Mikulenka provided a last-minute pass with six-hundred-grit sandpaper. Any remaining rough edges, knots, or errors are mine alone.

Finally, a number of people provided material or moral support. Sean Havey took my author photo and did the best he could with the subject matter. Erik Reyna and Dana Liebelson hosted me in Washington, D.C. Alison and Alan Henry hosted me in New Zealand. Leigh Brooks hosted me in Bristol, Florida. Ted St. George provided both a blow-up mattress and a car in Ann Arbor, Michigan. Jennifer Woo and Jens Driller were helpful throughout. Phuong Seltzer provided constant food and encouragement. Sandy and Jim St. George supported me in this, as they have in all my endeavors. Most of all, I would like to thank Tia Seltzer, who was there the whole time. Without you, I would be rootless.

NOTES

Introduction

3 *Swedish scientist*: Svante Arrhenius, *Worlds in the Making: The Evolution of the Universe*, trans. H. Borns (New York: Harper & Brothers, 1908), 61, 63.

3 *By the 1980s*: J. Hansen, D. Johnson, A. Lacis, S. Lebedeff, P. Lee, D. Rind, and G. Russell, "Climate Impact of Increasing Atmospheric Carbon Dioxide," *Science* 213, no. 4511 (August 28, 1981): 964–66.

5 *The oldest known trees*: W. E. Stein, F. Mannolini, L. V. Hernick, E. Landing, and C. M. Berry, "Giant Cladoxylopsid Trees Resolve the Enigma of the Earth's Earliest Forest Stumps at Gilboa," *Nature* 446, no. 7138 (2007): 904–7.

5 *another group of ancient trees*: C. Kevin Boyce and William A. DiMichele, "Arborescent Lycopsid Productivity and Lifespan: Constraining the Possibilities," *Review of Paleobotany and Palynology* 227 (2016): 97–110.

5 *More familiar trees*: Aljos Farjon, "Conifers of the World," *Kew Review* 73 (2018): 8.

5 *next 100 million years*: Colin Tudge, *The Tree: A Natural History of What Trees Are, How They Live, and Why They Matter* (New York: Three Rivers, 2005), 70–74.

6 *ruled the world*: E. C. Pielou, *The World of Northern Evergreens*, 2nd ed. (Ithaca: Cornell University Press, 2011), 67, 68.

7 *As the historian Jared Farmer*: Jared Farmer, *Trees in Paradise: A California History* (New York, W. W. Norton, 2013), xxxi–xxxii.

7 *represents endurance*: Asa Gray, *Scientific Papers of Asa Gray*, vol. 2, *Essays; Biographical Sketches, 1841–1886*, ed. Charles Sprague Sargent (Boston: Houghton Mifflin, 1889), 79.

1. They Seem to Be Immortal

11 *They are the world's biggest*: Nathan Stephenson, "Ecology and Management of Giant Sequoia Groves," *Sierra Nevada Ecosystem Project: Final Report to Congress*, vol. 2, *Assessments and Scientific Basis for Management Options* (Davis: University of California, Centers for Water and Wildland Resources, 1996), 1432.

11 *"Barring accidents"*: John Muir, *The Mountains of California* (1894; repr., Boston: Houghton Mifflin, 1916), 200.

12 *three years into a statewide drought*: Koren Nydick et al., "Leaf to Landscape Responses of Giant Sequoia to Hotter Drought: An Introduction and Synthesis for the Special Section," *Journal of Forest Ecology and Management* 419–20 (July 2018): 252.

12 *seventy-odd groves*: Stephenson, "Ecology and Management of Giant Sequoia Groves," 1434.

12 *narrow sliver*: Nydick, "Leaf," 252.

13 *first learned of sequoias*: Gary Lowe, *The Original Big Tree: History of the Big Tree Exhibit of 1853–1855* (Livermore, Calif.: printed by author, 2012), 13–23, 240.

13 *there had been rumors*: Ibid., 276.

13 *the Sylvan Mastodon*: Jared Farmer, *Trees in Paradise: The Botanical Conquest of California* (New York: W. W. Norton, 2013), 13.

13 *the Enormous Vegetable Production, the Big Cylinder*: Lowe, *Original Big Tree*, 289, 356.

13 *A few months later*: Ibid., 44–53.

13 *After the tree fell*: William Tweed, *King Sequoia: The Tree That Inspired a Nation, Created Our National Park System, and Changed the Way We Think About Nature* (Berkeley, Calif.: Heyday and Sierra College Press, 2016), 6.

14 *one magazine complained*: Lowe, *Original Big Tree*, 287.

14 *Said another*: "The Mammoth Tree Grove of California," *New York Herald*, December 17, 1855, 4.

14 *blotted out eastern skies*: Chih-Ming Hung et al., "Drastic Population Fluctuations Explain the Rapid Extinction of the Passenger Pigeon," *Proceedings of the National Academy of Sciences* 111, no. 29 (2014): 29.

14 *save most of them*: Stephenson, "Ecology and Management of Giant Sequoia Groves," 1435.

14 *Fire had always*: Ibid.

15 *"Have they had a career"*: Asa Gray, *Sequoia and Its History: An Address by Asa Gray* (Salem, [Mass.]: Salem Press, 1872), 6.

15 *prolific American botanist*: Walter Deane, "Asa Gray," *Bulletin of the Torrey Botanical Club* 15, no. 3 (March 2, 1888): 64–68.

15 *a trip out west*: Gray, *Sequoia and Its History*, 3–5.

15 *a young writer*: Tweed, *King Sequoia*, 62.

16 *where it was supposed to be*: Isaac Biberg, "The Economy of Nature" (Uppsala: master's thesis, 1749), in Carl von Linné, *Miscellaneous Tracts Relating to Natural History, Husbandry, and Physick: To Which Is Added the Calendar of Flora*, trans. Benjamin Stillingfleet (London: R. & J. Dodsley, 1762), 67.

16 *To further explain*: Carl von Linné, *Select Dissertations from the Amoenitates Academicae*, trans. F. J. Brand (1781; repr., New York: Arno, 1977), 88–94, 113–15, 126–27. In *Foundations of Biogeography: Classic Papers with Commentaries*, ed. Mark Lomolino, Dov Sax, and James Brown (Chicago: University of Chicago Press, 2004), 14, 15.

17 *the only animals shared*: Georges-Louis Leclerc, Compte de Buffon, *Natural History, General and Particular*, in *Foundations of Biogeography: Classic Papers with Commentaries*, ed. Mark Lomolino, Dov Sax, and James Brown (Chicago: University of Chicago Press, 2004), 16, 17.

17 *studying the flora*: Alexander von Humboldt and Aimé Bonpland, *Essay on the Geography of Plants*, trans. Sylvie Romanowksi (Chicago: University of Chicago Press, 2007), 68–69, 82–96.

17 *not once but many times*: Charles Lyell, *Principles of Geology: The Modern Changes of the Earth and Its Inhabitants*, 7th ed. (London: John Murray, 1847), 675.

18 *"The smallest grain"*: Charles Darwin and Alfred Russel Wallace, "On the Tendency of Species to Form Varieties," *Zoological Journal of the Linnean Society* 3 (August 20, 1858): 45–56.

18 *could, perhaps, be solved*: Gray, *Sequoia and Its History*, 7.

19 *mule named Brownie*: Tweed, *King Sequoia*, 64.

19 *looking for specifics*: John Muir, *Our National Parks* (Boston: Houghton Mifflin, 1902), 285.

19 *It is called biogeography*: James Brown, introduction to *Foundations of Biogeography: Classic Papers with Commentaries*, ed. Mark Lomolino, Dov Sax, and James Brown (Chicago: University of Chicago Press, 2004), 1.

19 *a recognized vocation*: Sydney Ross, "Scientist: The Story of a Word," *Annals of Science* 18, no. 2 (1962).

20 *"far strange land"*: John Muir, "The New Sequoia Forests of California," *Harper's Magazine* (November 1878).

20 *wayward sequoias*: John Muir, "On the Post-Glacial History of Sequoia," *Proceedings*

of the *American Association for the Advancement of Science 1876* (Washington, D.C.: American Association for the Advancement of Science, 1877), 5–9.

21 *"bounty of the clouds"*: Muir, *Our National Parks*, 324, 326.

21 *"passing away"*: John Muir, "God's First Temples," *Sacramento Daily Union* 1, no. 304 (1876).

21 *destruction to come*: Muir, "God's First Temples."

21 *In the late 1880s*: Hank Johnston, *They Felled the Redwoods: A Saga of Flumes and Rails in the High Sierra*, 3rd ed. (Glendale, Calif.: Trans-Anglo Books, 1973), 24, 25.

22 *a strange shade*: Ibid., 29–30, 36, 46, 129.

22 *Sanger held a party*: Ibid., 30–31.

22 *another of sequoias' ironies*: Douglas Piirto, "Wood Properties of Giant Sequoia: Properties and Unique Characteristics," *General Technical Report* (Berkeley, Calif.: Pacific Southwest Forest and Range Experiment Station, Forest Service 1986), 19–23.

22 *turned into little things*: Tweed, *King Sequoia*, 86.

23 *filed for bankruptcy*: Ibid., 83–85.

23 *When he died*: "John Muir, Aged Naturalist, Dead," *New York Times*, December 25, 1914.

23 *read schoolbooks*: John Muir, *The Story of My Boyhood and Youth* (Boston: Houghton Mifflin, 1913), 53–54, 83, 158–60.

24 *Martha*: Hung et al., "Drastic Population Fluctuations."

24 *buffalo herds*: Ben Potter et al., "History of Bison in North America," *American Bison: Status Survey and Conservation Guidelines 2010*, ed. C. Cormack Gates, Curtis Freese, Peter Gogan, and Mandy Kotzman (Gland, Switzerland: International Union for Conservation of Nature, 2010), 8.

24 *Californian grizzlies*: David Mattson and Troy Merrill, "Extirpations of Grizzly Bears in the Contiguous United States, 1850–2000," *Conservation Biology* 16, no. 4 (2002): 1123–36.

24 *western frontier*: Frederick Jackson Turner, *The Frontier in American History* (New York: Henry Holt, 1921): 1.

24 *begun to run on oil*: D. T. Armentano, "The Petroleum Industry: A Historical Study in Power," *Cato Journal* 1, no. 1 (1981): 53–59.

24 *Wilson had signed a bill*: Richard Lowitt, "The Hetch Hetchy Controversy, Phase II: The 1913 Senate Debate," *California History* 74, no. 2 (1995): 190–203.

24 *died of a broken heart*: Stephen Fox, *The American Conservation Movement: John Muir and His Legacy* (Boston: Little, Brown, 1981), 145–46.

24 *found among his papers*: John Muir, "Save the Redwoods," *Sierra Club Bulletin* 11, no. 1 (1920): 1–4.

24 *all but a few groves*: Tweed, *King Sequoia*, 195.

24 *upslope through chaparral*: Muir, *Our National Parks*, 307.

25 *the sequoias' seeds needed*: Muir, "New Sequoia Forests of California."

25 *Muir's view of fire*: Tweed, *King Sequoia*, 200.

25 *Peshtigo*: Eric Rutkow, *American Canopy: Trees, Forests, and the Making of a Nation* (New York: Scribner, 2012), 115–20.

25 *after the fires stopped*: Stephenson, "Ecology and Management of Giant Sequoia Groves," 1435.

25 *By the 1950s, the damage*: Tweed, *King Sequoia*, 207.

25 *known as the Leopold Report*: A. Starker Leopold et al., "Wildlife Management in the National Parks," Advisory Board on Wildlife Management Appointed by Secretary of the Interior Stewart Udall (March 4, 1963).

26 *managers started lighting fires*: Tweed, *King Sequoia*, 210.

26 *past fire frequency*: David Parsons, "Restoring Fire to Giant Sequoia Groves: What Have We Learned in 25 Years?" *Proceedings: Symposium on Fire in Wilderness and Park Management*, March 30–April 1, 1993, 256.

27 *sequoical State of the State*: Stephenson, "Ecology and Management of Giant Sequoia Groves."

28 *People mourn the diminishment*: Richard Hobbs, "Grieving for the Past and Hoping for the Future: Balancing Polarizing Perspectives in Conservation and Restoration," *Restoration Ecology* 21, no. 2 (2013): 145–48.

29 *"Niche" was originally*: G. Evelyn Hutchinson, *An Introduction to Population Ecology* (New Haven: Yale University Press, 1978).

29 *as an ecological concept*: Joseph Grinnell, "The Niche-Relationships of the California Thrasher," *Auk* 34, no. 4 (1917): 427–33.

29 *Elton offered another version*: Charles Elton, *Animal Ecology* (New York: Macmillan, 1927), 64.

30 *useful definition*: G. Evelyn Hutchinson, "Concluding Remarks," *Cold Spring Harbor Symposium on Quantitative Biology* 22 (1957): 415–27.

30 *Mediterranean climate*: Yanjun Su, "Emerging Stress and Relative Resiliency of Giant Sequoia Groves Experiencing Multi-Year Dry Periods in a Warming Climate," *Journal of Geophysical Research: Biogeoscience* 112, no. 11 (2017): 3063–75.

31 *warmer and drier*: Constance Millar and Wallace Woolfenden, "Ecosystems Past: Vegetation Prehistory," in *Ecosystems of California*, ed. Harold Mooney and Erika Zavalata (Berkeley: University of California Press, 2016), 146, 148.

31 *modern abundance*: Constance Millar and Linda Brubaker, "Climate Change and

Paleoecology: New Contexts for Restoration Ecology," in *Foundations of Restoration Ecology*, ed. Donald Falk, Margaret Palmer, and Joy Zedler (Washington, D.C.: Island Press, 2006), 327.

31 *little snow fell*: Soumaya Belmecheri et al., "Multi-Century Evaluation of Sierra Nevada Snowpack," *Nature Climate Change* 6, nos. 2–3 (2016): 1–2.

31 *at least a century*: Nydick et al., "Leaf to Landscape," 252, 254.

31 *bundles of straws*: M. T. Tyree and J. S. Sperry, "Vulnerability of Xylem to Cavitation and Embolism," *Annual Review of Plant Physiology and Plant Molecular Biology* 40 (June 1989): 19–38.

32 *a desperate move*: Anthony Ambrose et al., "Leaf- and Crown-Level Adjustments Help Sequoias Maintain Favorable Water Status During Severe Drought," *Forest Ecology and Management* 419–20 (2018): 257–67.

32 *confirm this conclusion*: Nathan Stephenson et al., "Patterns and Correlates of Giant Sequoia Foliage Dieback During California's 2012–2016 Hotter Drought," *Forest Ecology and Management* 419–20 (2018): 268–78.

32 *increased by about four degrees*: Yanjun Su et al., "Emerging Stress and Relative Resiliency of Giant Sequoia Groves," 6.

32 *the more water plants transpire*: Craig Allen et al., "A Global Overview of Drought and Heat-Induced Tree Mortality Reveals Emerging Climate Change Risks for Forests," *Forest Ecology and Management* 259, no. 4 (2010): 660–84.

33 *"Not one seed in a million"*: Muir, *Mountains of California*, 205.

33 *critical second look*: Eric Michael Johnson, "How John Muir's Brand of Conservation Led to the Decline of Yosemite," *Scientific American* (August 13, 2014).

34 *The Leopold Report is out, too*: "Revisiting Leopold: Resource Stewardship in the National Parks," National Park System Advisory Board Science Committee (August 25, 2012), 14.

34 *"continuously changing system"*: Ibid., 1.

34 *"would die before Sequoia"*: Muir, *Our National Parks*, 325.

35 *can live five hundred years*: Bohun B. Kinloch, *Sugar Pine: An American Wood*, USDA Forest Service, publication FS-257 (Washington, D.C.: U.S. Government Printing Office, 1984), 4.

35 *each year was rising*: Phillip van Mantgem et al., "Widespread Increase of Tree Mortality Rates in the Western United States," *Science* 323 (2009): 521–24.

35 *killed some 150 million trees*: Stephanie Gomes and Scott McLean, "Survey Finds 18 Million Trees Died in California in 2018" (CalFire, USDA Forest Service, February 11, 2018), https://www.fs.usda.gov/detail/catreemortality/toolkit/?cid=FSEPRD609121.

2. The Holocene

38 *Often their remains*: Jack Wolfe and Toshimasa Tanai, "The Miocene Seldovia Point Flora from the Kenai Group, Alaska," Geological Survey Professional Paper 1105 (Washington, D.C.: U.S. Government Printing Office, 1980).

38 *summing up a similar list*: Gray, "Forest Geography and Archaeology," *Scientific Papers of Asa Gray*, vol. 2, *Essays; Biographical Sketches*, ed. Charles Sprague Sargent (Boston: Houghton Mifflin, 1889), 228.

39 *"We infer the climate"*: Ibid., 227.

39 *as warm as New Jersey*: Asa Gray, *Sequoia and Its History: An Address by Asa Gray* (Salem, [Mass.]: Salem Press, 1872), 6.

39 *a startling claim*: Louis Agassiz, "Upon Glaciers, Moraines and Erratic Blocks: Being the Address Delivered at the Opening of the Helvetic Natural History Society Neuchâtel, on the 24th of July 1837, by Its President, M. L. Agassiz," *Edinburgh New Philosophical Journal* 24 (1838): 364–83.

39 *the flora of Japan*: Asa Gray, "The Flora of Japan," in *Scientific Papers of Asa Gray*, vol. 2, *Essays; Biographical Sketches 1841–1886*, ed. Charles Sprague Sargent (Boston: Houghton Mifflin, 1889), 138.

40 *a series of worldwide floods*: Georges Cuvier, *Essay on the Theory of the Earth* (New York: Kirk & Mercein, 1818), 38, 44.

41 *processes still visible*: James Hutton, *The Theory of the Earth* (Edinburgh: Royal Society of Edinburgh, 1788), 96.

41 *The current position of the continents*: Charles Lyell, *Principles of Geology: Being an Inquiry How Far the Former Changes of the Earth's Surface Are Referable to Causes Now in Operation*, 3rd ed. (London: John Murray, 1835), 176–82.

41 *sensitive to slight changes*: Mark Maslin, "How the Age of Ice Began," in *The Complete Ice Age*, ed. Brian Fagan (New York: Thames & Hudson, 2009), 50–83.

42 *These Milankovitch cycles*: E. C. Pielou, *After the Ice Age: The Return of Life to Glaciated North America* (Chicago: University of Chicago Press, 1991), 8–9.

42 *covered nearly all of Canada*: Ibid., 1–2.

42 *Then, around 18,000 years ago*: A. S. Dyke and V. K. Prest, "Paleogeography of Northern North America, 18,000–5,000 Years Ago," Canada Geological Survey Map 1703A, scale 1:12,500,000 (Ottawa: Geological Survey of Canada, 1987).

42 *Darwin compared the fossil record*: Charles Darwin, *On the Origin of Species by Means of Natural Selection* (London: John Murray 1859), 310–11.

46 *escape the fatal damage*: E. C. Pielou, *The World of Northern Evergreens*, 2nd ed. (Ithaca: Comstock Publishing, 2011), 6.

47 *a forest of deciduous trees*: Andrea H. Lloyd, Mary E. Edwards, Bruce P. Finney, Jason A. Lynch, Valerie Barber, and Nancy H. Bigelow, "Holocene Development of the Alaskan Boreal Forest," in *Alaska's Changing Boreal Forest*, ed. F. Stuart Chapin III, Mark W. Oswood, Keith Van Cleve, Leslie A. Viereck, David L. Verbyla, and Melissa C. Chapin (New York: Oxford University Press, 2006), 62–78.

48 *The early Holocene in Alaska*: Ibid., 333.

48 *the world's biggest forest*: Corey J. A. Bradshaw, Ian G. Warkentin, and Navjot S. Sodhi, "Urgent Preservation of Boreal Carbon Stocks and Biodiversity," *Trends in Ecology and Evolution* 24, no. 10 (2009): 541–48.

49 *Clement Reid*: Clement Reid, *The Origin of the British Flora* (London: Dalau & Co., 1899), 29.

49 *"something like a million years"*: Ibid., 25.

50 *white spruces had raced north*: J. C. Ritchie and G. M. MacDonald, "The Patterns of Post-glacial Spread of White Spruce," *Journal of Biogeography* 13, no. 6 (1986): 527–40.

50 *a mathematical framework*: James S. Clark, Chris Fastie, George Hurtt, Stephen T. Jackson, Carter Johnson, George A. King, Mark Lewis, Jason Lynch, Stephen Pacala, Colin Prentice, Eugene W. Schupp, Thompson Webb III, and Peter Wyckoff, "Reid's Paradox of Rapid Plant Migration," *BioScience* 48, no. 1 (1998): 13–24.

51 *"a large area of suitable soil"*: Eric Hultén, *Outline of the History of Arctic and Boreal Biota During the Quaternary: Their Evolution During and After the Glacial Period as Indicated by the Equiformal Progressive Areas of Present Plant Species* (1937; repr., New York: Wheldon & Wesley, 1971), 10–11.

51 *a German sea* Kapitän *debated*: Walter J. Eyerdam, "Botanical Collecting Rambles with Prof. Eric Hulten in the Aleutian Islands," *Madroño* 21, no. 4 (1971): 259–64.

52 *"It is often presumed"*: Hultén, *Outline*, 23.

52 *The bull's-eyes he'd found*: Ibid., 25.

52 *Among the few macrofossils*: G. D. Zazula, A. M. Telka, C. R. Harington, C. E. Schweger, and R. W. Mathewes, "New Spruce (*Picea* spp.) Macrofossils from Yukon Territory: Implications for Late Pleistocene Refugia in Eastern Beringia," *Arctic* 59, no. 4 (2006): 391–400.

53 *scientists have shown that both black and white spruces*: Lynn L. Anderson, Feng Sheng Hu, and Ken N. Page, "Phylogeographic History of White Spruce During the

Last Glacial Maximum: Uncovering Cryptic Refugia," *Journal of Heredity* (October 18, 2010); see also Sébastien Gérardi, Juan P. Jaramillo-Correa, Jean Beaulieu, and Jean Bousquet, "From Glacial Refugia to Modern Populations: New Assemblages of Organelle Genomes Generated by Differential Cytoplasmic Gene Flow in Transcontinental Black Spruce," *Molecular Ecology* 19, no. 23 (2010): 5265–80.

53 *modeled the climatic niches*: Jens-Christian Svenning and Flemming Skov, "Ice Age Legacies in the Geographical Distribution of Tree Species Richness in Europe," *Global Ecology and Biogeography* 16 (2007).

55 *The bristlecones are the world's oldest*: Eric Rutlow, *American Canopy: Trees, Forests, and the Making of a Nation* (New York: Scribner, 2012), 1–2.

55 *hoped to determine its age*: Ibid., 2–4.

57 *"Most scientists who have examined the evidence agree"*: John Imbrie and Katherine Palmer Imbrie, *Ice Ages: Solving the* Mystery (Cambridge, Mass.: Harvard University Press, 1979),177.

57 *Scientists have known for more than a century*: Charles C. Mann, *The Wizard and the Prophet: Two Remarkable Scientists and Their Dueling Visions to Shape Tomorrow's World* (New York: Alfred A. Knopf, 2018), 299–307.

57 *For more than fifty years*: "Restoring the Quality of Our Environment," *Report of the Environmental Pollution Panel* (Washington, D.C.: U.S. Government Printing Office, 1965), 113.

58 *"the measured responses of biosphere to climate"*: Pielou, *After the Ice Age*, 310.

58 *a new geological epoch*: Paul J. Crutzen and Eugene F. Stoermer, "The Anthropocene," *Global Change Newsletter* 41 (May 2000): 17–18.

58 *geologically arbitrary*: Stanley C. Finney and Lucy E. Edwards, "The 'Anthropocene' Epoch: Scientific Decision or Political Statement?" *GSA Today* 26, nos. 3–4 (2016).

58 *414 parts per million*: "Carbon Dioxide Levels Hit Record Peak in May," *NOAA Research News*, June 4, 2019, https://research.noaa.gov/article/ArtMID/587/ArticleID/2461/Carbon-dioxide-levels-hit-record-peak-in-May.

58 *280 parts per million*: Thure E. Curling, John M. Harris, Bruce J. MacFadden, Meave G. Leakey, Jay Quade, Vera Eisenmann, and James R. Ehleringer, "Global Vegetation Change Through the Miocene/Pliocene Boundary," *Nature* 189, no. 6647 (1997): 153–58.

58 *the mid-Pliocene*: Will Steffen, Johan Rockström, Katherine Richardson, Timothy M. Lenton, Carl Folke, Diana Liverman, Colin P. Summerhayes, Anthony D. Barnosky, Sarah E. Cornell, Michel Crucifixi, Jonathan F. Donges, Ingo Fetzer, Steven J. Lade, Marten Scheffer, Ricarda Winkelmann, and Hans Joachim Schellnhuber,

"Trajectories of the Earth System in the Anthropocene," *Proceedings of the National Academy of Sciences* 115, no. 33 (2018): 8252–59.

58 *the Eocene epoch*: Gavin L. Foster, Dana L. Royer, and Daniel J. Lunt, "Future Climate Forcing Potentially Without Precedent in the Last 420 Million Years," *Nature Communications* 8, no. 14845 (2017).

58 *when metasequoias grew in the Arctic*: Ralph W. Chaney, "A Revision of Fossil Sequoia and Taxodium in Western North America Based on the Recent Discovery of Metasequoia," *Transactions of the American Philosophical Society* 40, part 3 (February 1951): 173.

58 *violent periods of change in the past*: Steffen et al., "Trajectories of the Earth System"; see the supporting information.

58 *"climate velocity"*: Scott R. Loarie, Philip B. Duffy, Healy Hamilton, Gregory P. Asner, Christopher B. Field, and David D. Ackerly, "The Velocity of Climate Change," *Nature* 462, no. 7276 (2009): 1052–57.

58 *five miles per year*: Richard T. Corlett and David A. Westcott, "Will Plant Movements Keep Up with Climate Change?" *Trends in Ecology and Evolution* 28, no. 8 (2013): 482–88.

59 *hit its zenith*: Maslin, "How the Age of Ice Began," 64.

59 *subcontinental extension of Asia*: John F. Hoffecker, Scott A. Elias, Dennis H. O'Rourke, G. Richard Scott, and Nancy Bigelow, "Beringia and the Global Dispersal of Modern Humans," *Evolutionary Anthropology* 25 (2016): 64–78.

59 *"which I shall hereinafter call Beringia"*: Hultén, *Outline*, 34.

59 *topic of debate*: Michael R. Waters, "Late Pleistocene Exploration and Settlement of the Americas by Modern Humans," *Science* 365, no. 5447 (2019).

59 *32,000 years ago*: Richard Vachula, Yongsong Huang, William Longo, Sylvia Dee, William Daniels, and James Russell, "Evidence of Ice Age Humans in Eastern Beringia Suggests Early Migration to North America," *Quaternary Science Reviews* 205 (2019).

60 *large grazing animals*: B. Shapiro, A. J. Drummond, A. Rambaut, et al., "Rise and Fall of the Beringian Steppe Bison," *Science* 306 (2004): 1561–65.

61 *the early Beringians had spread out*: Waters, "Late Pleistocene Exploration."

61 *paleoecologist Paul Martin*: Paul Martin, "Africa and the Pleistocene Overkill," *Nature* 212, no. 5060 (1966): 339–42.

61 *"thought only of steaks"*: Aldo Leopold, *A Sand County Almanac and Other Writings on Ecology and Conservation* (1949; repr., New York: Literary Classics of the United States, 2013), 98.

3. How Monterey Pine Became Radiata and Other Stories

63 *became farmer-gardeners*: Jared Diamond, *Guns, Germs, and Steel: The Fates of Human Societies* (New York: W. W. Norton, 1999), 101.

64 *Jared Diamond wrote*: Ibid., 104.

64 *had it pretty good*: Yuval Noah Harari, *Sapiens: A Brief History of Humankind* (New York: Harper, 2011).

65 *Queen Hatshepsut*: John W. Turnbull, "Tree Domestication and the History of Plantations," in *The Role of Food, Agriculture, Forestry and Fisheries in Human Nutrition*, ed. Victor R. Squires (Oxford: EOLSS, 2009), 48–75.

65 *A mural at the Deir el-Bahari*: Jules Janick, "Plant Exploration: From Queen Hatshepsut to Sir Joseph Banks," *Journal of the American Society for Horticultural Science* 42, no. 2 (2007): 191–96.

65 *The Romans took a similar interest*: Carolyn Fry, *The Plant Hunters: The Adventures of the World's Greatest Botanical Explorers* (Chicago: University of Chicago Press, 2013), 13–15.

65 *wine grapes*: Cornelius Tacitus, *Tacitus' Germania and Agricola*, trans. E. A. Beaty (New York: Translation Publishing, 1933), 21.

65 *spice markets of Asia*: Fry, *Plant Hunters*, 17.

65 *"being restless and industrious"*: Alexander von Humboldt and Aimé Bonpland, *Essay on the Geography of Plants*, trans. Sylvie Romanowksi (Chicago: University of Chicago Press, 2007), 71.

66 *David Douglas was born*: William Hooker, "A Brief Memoir of the Life of Mr. David Douglas, with Extracts from His Letters," in *Companion to the Botanical Magazine*, vol. 2 (London: Samuel Curtis, 1836), 79–82.

66 *Europe's age of exploration*: Fry, *Plant Hunters*, 40–52.

67 *Douglas was twenty-four*: David Douglas, *Journal Kept by David Douglas During His Travels in North America, 1823–1827* (London: William Wesley & Son, 1914), 2–30.

67 *up an oak tree*: Ibid., 14–16.

67 *the trip was a success*: Ibid., 30–31.

67 *to the Pacific Northwest*: Ibid., 51–76.

67 *rusty nail [. . .] enormous rats*: Ibid., 93, 121, 128, 110, 115.

68 *popular in high society*: Ibid., 142.

68 *"manufacture Pines at my pleasure"*: Ibid., 152.

68 *"thoroughly unpretentious credentials"*: Peter B. Lavery and Donald J. Mead, "*Pinus radiata:* A Narrow Endemic from North America Takes On the World," *Ecology and*

Biogeography of Pinus, ed. David M. Richardson (Cambridge: Cambridge University Press, 1998), 432–60.

69 *Like the giant sequoia and the Florida torreya*: David T. Foster-Bates, Rita Dalessio, Nicole Nedeff, and Joyce Stevens, *Coastal California's Living Legacy: The Monterey Pine Forest* (Carmel-by-the-Sea, Calif.: Pine Nut Press, 2011), 3.

69 *on the Columbia River*: Hooker, "Brief Memoir," 156.

69 *from Oahu*: Ibid., 156–61.

69 *found dead on the flanks of Mauna Kea*: Ibid., 178.

70 *wrote one* Mrs. *Lyman*: Jean Greenwell, "Kaluakuaka Revisited: The Death of David Douglas in Hawaii," *Hawaiian Journal of History*, 22 (1988): 147–69.

70 *By the time of his death*: Hooker, "Brief Memoir," 182.

70 *But one tree Douglas would not get to name*: Lavery and Mead, "*Pinus radiata*," 433.

71 *almost entirely forested*: Rowland D. Burdon, William J. Libby, and Alan G. Brown, *Domestication of Radiata Pine* (Cham, Switzerland: Springer, 2017): 17.

71 *Captain James Cook arrived*: Ibid., 17.

71 *Joseph Banks was a naturalist*: Joseph Banks, *Journal of the Right Hon. Sir Joseph Banks Bart., K.B., P.R.S.: During Captain Cook's First Voyage in H.M.S.* Endeavour *in 1768–71 to Terra del Fuego, Otahite, New Zealand, Australia, the Dutch East Indies, Etc.*, ed. Joseph Hooker (London: Macmillan, 1896), 223–24.

71 *First came whalers and sailors*: Burdon, Libby, and Brown, *Domestication*, 17–18.

72 *several times noted*: Joseph Banks, *Journal*, 228.

72 *advent of refrigerated shipping*: New Zealand History, "First Frozen Meat Shipment Leaves New Zealand, 15 February 1882," Ministry for Culture and Heritage, 2019, https://nzhistory.govt.nz/first-shipment-of-frozen-meat-leaves-nz.

72 *there arose a new worry*: Michael Roche, "Exotic Forestry," *Te Ara: The Encyclopedia of New Zealand* (November 24, 2008), https://teara.govt.nz/en/exotic-forestry.

72 *Among the contestants*: Burdon, Libby, and Brown, *Domestication*, 18.

72 *a report on the nation's forests*: Royal Commission on Forestry, *Royal Commission on Forestry (Report of the), Together with Minutes of Proceedings and of Evidence* (New Zealand, 1913).

73 *This was a surprise*: Chris Hegan, "Radiata, Prince of Pines," *New Zealand Geographic* no. 020 (October–December 1993), https://www.nzgeo.com/stories/radiata-prince-of-pines.

73 *"every variety of soil"*: Royal Commission on Forestry, *Royal Commission*, 41.

74 *"improved variety of the species"*: Ibid., 42.

75 *two parts to domestication*: Burdon, Libby, and Brown, *Domestication*, 3.

75 *Foresters across the Southern Hemisphere*: Ibid., 10–11.

75 *the seeds David Douglas collected*: Ibid., 22.

76 *five hundred times greater*: Ibid., 10.

76 *Rarity precedes extinction*: Charles Darwin, *Journal of Researches into the Natural History and Geology of the Countries Visited During the Voyage of H.M.S.* Beagle *Round the World, Under the Command of Capt. Fitz Roy, R.N.*, 2nd ed. (1845; repr., New York: D. Appleton, 1871), 176, 194.

77 *sudden, apocalyptic upheavals*: Georges Cuvier, *Essay on the Theory of the Earth* (1796; repr., New York: Kirk & Mercein, 1818), 35–38.

77 *Asa Gray echoed Darwin*: Asa Gray, *Sequoia and Its History: An Address by Asa Gray* (Salem, [Mass.]: Salem Press, 1872), 7–8.

77 *Most early conservationists*: Richard Ladle and Robert Whittaker, *Conservation Biogeography* (Oxford: Wiley-Blackwell, 2011), 5.

78 *John and William Bartram*: Helen Gere Cruikshank, ed., *John and William Bartram's America: Selections from the Writings of the Philadelphia Naturalists* (New York: Devin-Adair, 1957), 95–98.

79 *People argued*: Vernon H. Heywood, "The Future of Plant Conservation and the Role of Botanic Gardens," *Plant Diversity* 39 (2017): 309–13.

80 *Torrey's student*: Asa Gray, "A Pilgrimage to Torreya," *Scientific Papers of Asa Gray*, vol. 2, *Essays; Biographical Sketches*, ed. Charles Sprague Sargent (Boston: Houghton Mifflin, 1889), 195.

80 *letter to the editor*: R. K. Godfrey and Herman Kurz, "The Florida Torreya: Destined for Extinction," *Science* 136, no. 3519 (1962): 900–902.

81 *Florida torreya be listed*: Fish and Wildlife Service, "Proposal to Determine *Torreya taxifolia* (Florida torreya) as an Endangered Species," *Federal Register* 46, no. 68 (1983): 15168–71.

82 *in their 1981 book*: Paul R. Ehrlich and Anne H. Ehrlich, *Extinction: The Causes and Consequences of the Disappearance of Species* (New York: Random House, 1981): xi–xiv.

83 *where pollen can float*: Rob Nicholson, "Chasing Ghosts," *Natural History* 99, no. 12 (1990): 8–13.

83 *a third option*: Constance Millar, "Reconsidering the Conservation of Monterey Pine," *Fremontia* 26, no. 3 (1998): 12–16.

84 *Robert Peters and Joan Darling*: Robert L. Peters and Joan D. S. Darling, "The Greenhouse Effect and Nature Reserves," *BioScience* 35, no. 11 (1985): 707–17.

84 *Margaret Davis*: Margaret B. Davis, "Lags in Vegetation Response to Greenhouse Warming," *Climate Change* 15 (1989): 75–82.

84 *Richard Primack and S. L. Miao*: Richard B. Primack and S. L. Miao, "Dispersal Can Limit Local Plant Distribution," *Conservation Biology* 6, no. 4 (1992): 513–19.

85 *"ecological anachronisms"*: Connie Barlow, *The Ghosts of Evolution: Nonsensical Fruit, Missing Partners, and Other Ecological Anachronisms* (New York: Basic Books, 2000).

85 *"present limited southern quarters"*: Asa Gray, *Sequoia and Its History*, 10.

85 *"a northern plant"*: John M. Coulter and W. J. G. Land, "Gametophytes and Embryo of *Torreya taxifolia*," *Botanical Gazette* 39, no. 3 (1905): 161–78.

85 *1986 recovery plan*: Fish and Wildlife Service, "Florida Torreya (*Torreya taxifolia*) Recovery Plan" (Atlanta, Ga.: U.S. Department of the Interior, Fish and Wildlife Service, Southeast Region, 1986).

85 *"It wants to head north"*: Connie Barlow, "Anachronistic Fruit and the Ghosts Who Haunt Them," *Arnoldia*, 61, no. 2 (2001): 14–21.

85 *paleobotanist Hazel Delcourt*: Hazel Delcourt, *Forests in Peril: Tracking Deciduous Trees from Ice-Age Refuges into the Greenhouse World* (Newark, Ohio: McDonald & Woodward Publishing, 2002), 207.

86 *paleoecologist Paul Martin*: Barlow, *Ghosts of Evolution*.

87 *the* Yes *opinion*: Connie Barlow and Paul S. Martin, "Bring *Torreya taxifolia* North—Now," *Wild Earth* (Fall–Winter 2004–5): 72, 74–77.

87 *Mark Schwartz anticipated*: Mark W. Schwartz, "Assessing the Ability of Plants to Respond to Climatic Change Through Distribution Shifts," in *Proceedings, 1995 Meeting of the Northern Global Change Program, March 14–16, 1995, Pittsburgh, PA.: Introduction/Environmental Change*, ed. John Hom, Richard Birdsey, and Kelly O'Brian (Radnor, Pa.: USDA Forest Service, 1996), 184–91.

4. Kiss Your Ash Good-Bye

89 *Étienne Léopold Trouvelot*: Andrew Liebhold, Victor Mastro, and Paul W. Schaefer, "Learning from the Legacy of Léopold Trouvelot," *Bulletin of the Entomological Society of America* 35, no. 2 (Summer 1989): 20–22.

89 *"that crawling repulsive creature"*: L. Trouvelot, "The American Silkworm," *American Naturalist* 2 (1868): 30–38.

89 *the eggs of the European gypsy moth*: Liebhold, Mastro, and Schaefer, "Learning from the Legacy."

90 *an 1896 report*: Edward H. Forbush and Charles H. Fernald, *The Gypsy Moth* (Boston: Wright & Potter Printing, 1896), 4.

90 *For a decade*: Liebhold, Mastro, and Schaefer, "Learning from the Legacy."

90 *the caterpillars reappeared*: Forbush and Fernald, *Gypsy Moth*, 15, 14, 9, 17.

91 *They continue to spread*: Bruce W. Kauffman, Wayne K. Clatterbuck, Andrew M. Liebhold, and David R. Coyle, "Gypsy Moth in the Southeastern U.S.: Biology, Ecology, and Forest Management Strategies," *Southern Regional Extension Forestry* (February 2017): 1–10.

91 *"wicked bad here"*: Gregory B. Hladky, "Gypsy Moth Damage in Eastern Connecticut Is Widespread," *Hartford Courant* (June 30, 2017), http://www.courant.com/news/connecticut/hc-gypsy-moth-deforestation-injuries-20170626-story.html.

91 *"New Pangaea"*: Elizabeth Kolbert, *The Sixth Extinction: An Unnatural History* (New York: Henry Holt, 2014), 193–216.

91 *Pangaea split*: Sarah McIntyre, Charles Lineweaver, Colin Groves, and Aditya Chopra, "Global Biogeography Since Pangaea," *Proceedings of the Royal Society B* 284, no. 1856 (2017): 284.

92 *"all the rage"*: Peter Coates, *American Perceptions of Immigrant and Invasive Species: Strangers on the Land* (Berkeley: University of California Press, 2006), 4, 35.

92 *Even governments participated*: Mark A. Davis, *Invasion Biology* (New York: Oxford University Press, 2009), 7.

92 *Swedish naturalist Pehr Kalm*: Peter Kalm, *Travels into North America*, trans. John Reinhold Forster (London, 1771), 9, 11, 46–47, 13, 16.

93 *pushed the old inhabitants out*: Charles Darwin, *On the Origin of Species by Means of Natural Selection* (London: John Murray, 1859), 66, 337.

93 *more than mere nuisance*: George Perkins Marsh, *The Earth as Modified by Human Action: A Last Revision of "Man and Nature"* (1864; repr., New York: Charles Scribner's Sons, 1898), 55–56.

93 *the zoo's chestnut trees*: Susan Freinkel, *American Chestnut: The Life, Death, and Rebirth of a Perfect Tree* (Berkeley: University of California Press, 2007), 15.

94 *"one out of every four trees"*: Charles C. Mann, *1491: New Revelations of the Americas Before Columbus*, 2nd ed. (New York: Vintage, 2011), 302.

94 *Dutch elm disease*: Thomas J. Campanella, *Republic of Shade: New England and the American Elm* (New Haven: Yale University Press, 2003), 1.

94 *American attitudes*: Coates, *American Perceptions*, 31, 38.

94 *International trade ballooned*: Gary M. Lovett, Marissa Weiss, Andrew M. Liebhold, Thomas P. Holmes, Brian Leung, Kathy Fallon Lambert, David A. Orwig, Faith T.

Campbell, Jonathan Rosenthal, Deborah G. McCullough, Radka Wildova, Matthew P. Ayres, Charles D. Canham, David R. Foster, Shannon L. LaDeau, and Troy Weldy, "Nonnative Forest Insects and Pathogens in the United States: Impacts and Policy Options," *Ecological Applications* 26, no. 5 (2016): 1437–55.

95 *"It is not just nuclear bombs"*: Charles Elton, *The Ecology of Invasions by Animals and Plants* (1958; repr., Chicago: University of Chicago Press, 2000), 15, 111, 30–31.

96 *sixteen species of ashes*: Kamal J. K. Gandhi and Daniel A. Herms, "North American Arthropods at Risk Due to Widespread *Fraxinus* Mortality Caused by the Alien Emerald Ash Borer," *Biological Invasions* 12, no. 6 (June 2010): 1839–46.

96 *Ohio alone held nearly four billion ash trees*: Ibid.

97 *The haft of Achilles's spear*: Robert Penn, *The Man Who Made Things Out of Trees: The Ash in Human Culture and History* (New York: W. W. Norton, 2016), 9.

97 *Some three hundred million ash trees*: Animal and Plant Health Inspection Service, USDA, "Proposed Release of Three Parasitoids for the Biological Control of the Emerald Ash Borer (*Agrilus planipennis*) in the Continental United States, Environmental Assessment" (April 2, 2007), 19.

97 *large genus of beetles*: M. Lourdes Chamorro, Eduard Jendek, Robert A. Haack, Toby R. Petrice, Norman E. Woodley, Alexander S. Konstantinov, Mark G. Volkovitsh, Xing-Ke Yang, Vasily V. Grebennikov, and Steven W. Lingafelter, *Illustrated Guide to the Emerald Ash Borer* Agrilus planipennis *Fairmaire and Related Species (Coleoptera, Buprestidae)* (Sofia, Bulgaria: Pensoft Publishers, 2015), 22.

99 *By July 2002*: Daniel A. Herms and Deborah G. McCullough, "Emerald Ash Borer Invasion of North America: History, Biology, Ecology, Impacts, and Management," *Annual Review of Entomology* 59 (October 9, 2013): 13–30.

99 *Chinese forestry textbook*: Chengming Yu, "*Agrilus marcopoli* Odenberger (Coleoptera: Buprestidae)," in *Forest Insects of China*, trans. Houping Liu, 2nd ed. (Beijing: China Forestry Publishing House, 1992), 400–401.

100 *A state report from the late 1920s*: William N. Sparhawk and Warren D. Brush, *The Economic Aspects of Forest Destruction in Northern Michigan* (Washington, D.C.: U.S. Government Printing Office, 1930).

100 *back up to roughly half*: Scott A. Pugh, Mark H. Hansen, Lawrence D. Pedersen, Douglas C. Heym, Brett J. Butler, Susan J. Crocker, Dacia Meneguzzo, Charles H. Perry, David E. Haugen, Christopher W. Woodall, and Ed Jepsen, *Michigan's Forests*, Resource Bulletin NRS-34 (Newtown Square, Pa.: USDA Forest Service, 2004).

101 *enacted a quarantine*: Robert A. Haack, Yuri Baranchikov, Leah S. Bauer, and Therese M. Poland, "Emerald Ash Borer Biology and Invasion History," in *Biology and Control*

of *Emerald Ash Borer*, ed. Roy G. Van Driesche and Richard C. Reardon (Morgantown, W.Va.: Forest Health Technology Enterprise Team, March 2015), 5–7.

103 *emerald ash borer invasion*: Nathan W. Siegert, Deborah G. McCullough, Andrew M. Liebhold, and Frank W. Telewski, "Dendrochronological Reconstruction of the Epicentre and Early Spread of Emerald Ash Borer in North America," *Diversity and Distributions* 20 (2014): 847–58.

104 *entomologist Sandy Liebhold*: Andrew Liebhold, "Population Processes During Establishment and Spread of Invading Species: Implications for Survey and Detection Programs," in *Detecting and Monitoring of Invasive Species: Plant Health Conference 2000* (Raleigh, N.C.: U.S. Department of Agriculture, 2000).

105 *between 1970 and 2010*: United Nations Conference on Trade and Development, *Review of Marine Transport 2018* (New York: United Nations Publications, 2018), 5, x.

105 *another 66 million tons*: International Air Transport Association, "IATA Cargo Strategy" (February 2018), 5, https://www.iata.org/whatwedo/cargo/Documents/cargo-strategy.pdf.

105 *planes also carried*: International Air Transport Association, "Traveler Numbers Reach New Heights" (September 6, 2018), https://www.iata.org/pressroom/pr/Pages/2018–09–06–01.aspx.

105 *as many as 3 billion live plants*: Andrew M. Liebhold, Eckehard G. Brockerhoff, Lynn J. Garrett, Jennifer L. Parke, and Kerry O. Britton, "Live Plant Imports: The Major Pathway for Forest Insect and Pathogen Invasions of the US," *Frontiers in Ecology and the Environment* 10, no. 3 (2012): 135–43.

105 *rough rule of tens*: Mark Williams and Alastair Fitter, "The Varying Success of Invaders," *Ecology* 77, no. 6 (1996): 1661–66.

106 *that makes an invader*: Daniel Simberloff, *Invasive Species: What Everyone Needs to Know* (New York: Oxford University Press, 2013), 145–59.

107 *Asa Gray wrote in 1876*: Asa Gray, "The Pertinacity and Predominance of Weeds," in *Scientific Papers of Asa Gray*, vol. 2, *Essays; Biographical Sketches, 1841–1886*, ed. Charles Sprague Sargent (Boston: Houghton Mifflin, 1889), 236.

107 *ancient temperate forest*: Michael Donoghue and Stephen Smith, "Patterns in the Assembly of Temperate Forests Around the Northern Hemisphere," *Philosophical Transactions of the Royal Society B* 359 (2004): 1633–44.

109 *"No one is likely to get into New Zealand"*: Elton, *Ecology of Invasions*, 111.

109 *more than five million species of insects*: Nigel E. Stork, "How Many Species of Insects and Other Terrestrial Arthropods Are There on Earth?" *Annual Review of Entomology* 63, no. 1 (2018): 31–45.

109 *untold millions more fungi*: Meredith Blackwell, "The Fungi: 1, 2, 3 . . . 5.1 million Species?" *American Journal of Botany* 98, no. 3 (2011): 426–38.

109 *bacteria, and viruses*: Daniel Dykhuizen, "Species Numbers in Bacteria," *Proceedings of the California Academy of Sciences* 56, no. 6 (2005): 62–71.

110 *"a long record of tinkering"*: Mark Schwartz, "Conservationists Should Not Move *Torreya taxifolia*," *Wild Earth* (Fall–Winter 2004–5): 73, 77–79.

111 *researchers led by Jason Smith*: J. A. Smith, K. O'Donnell, L. L. Mount, K. Shin, A. Trulock, T. Spector, J. Cruse-Sanders, and R. Determann, "A Novel *Fusarium* Species Causes a Canker Disease of the Critically Endangered Conifer, *Torreya taxifolia*," *Plant Disease* 95 (2011): 633–39.

111 *a follow-up paper*: Takayuki Aoki, Jason A. Smith, Lacey L. Mount, David M. Geiser, and Kerry O'Donnell, "*Fusarium torreyae* sp. nov., a Pathogen Causing Canker Disease of Florida torreya (*Torreya taxifolia*), a Critically Endangered Conifer Restricted to Northern Florida and Southwestern Georgia," *Mycologia* 105, no. 2 (2013): 312–19.

111 *reported the results*: Aaron J. Trulock, "Host Range and Biology of *Fusarium torreyae* (sp. nov.), Causal Agent of Canker Disease of Florida Torreya (*Torreya taxifolia* Arn)" (master's thesis, University of Florida, 2012), 17–24, 26.

112 *a dangerous precedent*: Mark Schwartz, "Conservationists Should Not Move," 78.

113 *Fossil evidence of genus* Torreya: Connie Barlow and Paul S. Martin, "Bring *Torreya taxifolia* North—Now," *Wild Earth* (Fall–Winter 2004–5): 75.

114 *one of the most destructive invaders*: Herms and McCullough, "Emerald Ash Borer Invasion of North America," 17.

114 *planted North American ashes*: Ibid., 23.

114 *European ash*: Deepa S. Pureswaran and Therese M. Poland, "Host Selection and Feeding Preference of *Agrilus planipennis* (Coleoptera: Buprestidae) on Ash (*Fraxinus* spp.)," *Environmental Entomology* 38, no. 3 (2009): 757–65.

114 *at risk of extinction*: International Union of Concerned Scientists, "Once-Abundant Ash Tree and Antelope Species Face Extinction—IUCN Red List" (September 14, 2017), https://www.iucn.org/news/secretariat/201709/once-abundant-ash-tree-and-antelope-species-face-extinction-%E2%80%93-iucn-red-list.

114 *stricter controls on wooden pallets*: Robert A. Haack, Kerry O. Britton, Eckehard G. Brockerhoff, Joseph F. Cavey, Lynn J. Garrett, Mark Kimberly, Frank Lowenstein, Amelia Nuding, Lars J. Olson, James Turner, and Katheryn N. Vasilaky, "Effectiveness of the International Phytosanitary Standard ISPM No. 15 on Reducing Wood Borer Infestation Rates in Wood Packaging Material Entering the United States," *PLOS One* 9, no. 5 (2014).

115 *"aroused so much discussion"*: Christina Devorshak, "History of Plant Quarantine and the Use of Risk Analysis," in *Plant Pest Risk Analysis: Concepts and Application*, ed. Christina Devorshak (Oxfordshire: CABI, 2012), 19–28.

115 *remained relatively steady*: Lovett et al., "Nonnative Forest Insects and Pathogens in the United States."

116 *as damaging as or worse*: Juliann E. Aukema, Brian Leung, Kent Kovacs, Corey Chivers, Kerry O. Britton, Jeffrey Englin, Susan J. Frankel, Robert G. Haight, Thomas P. Holmes, Andrew M. Liebhold, Deborah G. McCullough, and Betsy Von Holle, "Economic Impacts of Non-native Forest Insects in the Continental United States," *PLOS One* 6, no. 9 (2011).

5. Counterpest

118 *Gaps appeared*: Kamal J. K. Gandhi and Daniel A. Herms, "Direct and Indirect Effects of Alien Insect Herbivores on Ecological Processes and Interactions in Forests of Eastern North America," *Biological Invasions* 12 (2010): 389–405.

119 *The increase in sunlight*: Kamal J. K. Gandhi, Annemarie Smith, Diane M. Hartzler, and Daniel A. Herms, "Indirect Effects of Emerald Ash Borer-Induced Ash Mortality and Canopy Gap Formation on Epigaeic Beetles," *Environmental Entomology* 43, no. 3 (2014): 546–55.

119 *earthworms increased in abundance*: Michael D. Ulyshen, Wendy S. Klooster, William T. Barrington, and Daniel A. Herms, "Impacts of Emerald Ash Borer-Induced Tree Mortality on Leaf Litter Arthropods and Exotic Earthworms," *Pedobiologia: International Journal of Soil Biology* 54 (2011): 261–65.

119 *Worst hit were the arthropods*: Kamal J. K. Gandhi and Daniel A. Herms, "North American Arthropods at Risk Due to Widespread Fraxinus Mortality Caused by the Alien Emerald Ash Borer," *Biological Invasions* 12, no. 6 (2010): 1839–46.

119 *replacing dead ash trees*: T. Davis Sydnor, Matthew Bumgardner, and Sakthi Subburayalu, "Community Ash Densities and Economic Impact Potential of Emerald Ash Borer (*Agrilus planipennis*) in Four Midwestern States," *Arboriculture & Urban Forestry* 37, no. 2 (2011): 84–89.

119 *"additional 6,113 deaths"*: Geoffrey H. Donovan, David T. Butry, Yvonne L. Michael, Jeffrey P. Prestemon, Andrew M. Liebhold, Demetrios Gatziolis, and Megan Y. Mao, "The Relationship Between Trees and Human Health: Evidence from the Spread of Emerald Ash Borer," *American Journal of Preventative Medicine* 44, no. 2 (2013): 139–45.

119 *increase in crime*: Michelle C. Kondo, SeungHoon Han, Geoffrey H. Donovan, and

John M. MacDonald, "The Association Between Urban Trees and Crime: Evidence from the Spread of the Emerald Ash Borer in Cincinnati," *Landscape and Urban Planning* 157 (2017): 193–99.

120 *"The equation of animal and vegetable life"*: George Perkins Marsh, *The Earth as Modified by Human Action: A Last Revision of "Man and Nature"* (1864; repr., New York: Charles Scribner's Sons, 1898), 143.

121 *the botanist Ji Han*: H. T. Huang and Pei Yang, "The Ancient Cultured Citrus Ant: A Tropical Ant Used to Control Insect Pests in Southern China," *BioScience* 37, no. 9 (1987): 665–71.

121 *"noxious foreign species"*: Marsh, *Man and Nature*, 134.

122 *produced only two seeds*: Charles Darwin, *On the Origin of Species by Means of Natural Selection* (London: John Murray 1859), 66.

122 *"leaves no poisonous residues"*: Rachel Carson, *Silent Spring* (Boston: Houghton Mifflin, 1962), 292.

122 *"very wide proportions"*: Charles Elton, *Ecology of Invasions by Animals and Plants* (1958; repr., Chicago: University of Chicago Press, 2000), 131, 83.

123 *Houping Liu traveled*: Houping Liu, Leah S. Bauer, Ruitong Gao, Tonghai Zhao, Toby R. Petrice, and Robert A. Haack, "Exploratory Survey for the Emerald Ash Borer, *Agrilus planipennis* (Coleoptera: Buprestidae), and Its Natural Enemies in China," *Great Lakes Entomologist* 36, no. 3–4 (2003): 191–204.

125 *numerous species of native Hawaiian snails*: R. H. Cowie, "Can Snails Ever Be Effective and Safe Biocontrol Agents?" *International Journal of Pest Management* 47, no. 1 (2001): 23–40.

125 *counterpests of the gypsy moth*: Ronald M. Weseloh, "People and the Gypsy Moth: A Story of Human Interactions with an Invasive Species," *American Entomologist* 49, no. 3 (2003): 180–90.

125 *people introduced mongooses*: A. Barun, C. C. Hanson, K. J. Campbell, and D. Simberloff, "A Review of Small Indian Mongoose Management and Eradications on Islands," in *Island Invasives: Eradication and Management*, ed. C. R. Veitch, M. N. Clout, and D. R. Towns (Gland, Switzerland: IUCN, 2011), 17–25.

126 *USDA scientists had to be sure*: Animal and Plant Health Inspection Service, USDA, "Proposed Release of Three Parasitoids for the Biological Control of the Emerald Ash Borer (*Agrilus planipennis*) in the Continental United States, Environmental Assessment" (April 2, 2007).

128 *killed something like 20 percent*: Leah S. Bauer, Jian J. Duan, Juli R. Gould, and Roy Van Driesche, "Progress in the Classical Biological Control of *Agrilus planipen-*

nis Fairmaire (Coleoptera: Buprestidae) in North America," *Canadian Entomologist* 147 (2015): 300–317.

128 *killed 99.7 percent of mature ash trees*: Wendy S. Klooster, Daniel A. Herms, Kathleen S. Knight, Catherine P. Herms, Deborah G. McCullough, Annemarie Smith, Kamal J. K. Gandhi, and John Cardina, "Ash (*Fraxinus* spp.) Mortality, Regeneration, and Seed Bank Dynamics in Mixed Hardwood Forests Following Invasion by Emerald Ash Borer (*Agrilus planipennis*)," *Biological Invasions* 16 (2014): 859–73.

128 *These are rare numbers*: Charles C. Mann, *1491: New Revelations of the Americas Before Columbus*, 2nd ed. (New York: Vintage, 2011), 108–21.

128 *after Europeans arrived*: John Lindo, Emilia Huerta-Sánchez, Shigeki Nakagome, Morten Rasmussen, Barbara Petzelt, Joycelynn Mitchell, Jerome Cybulski, Eske Willerslev, Michael DeGiorgio, and Ripan Malhi, "A Time Transect of Exomes from a Native American Population Before and After European Contact," *Nature Communications* 13175, no. 7 (2016).

129 *As Peter Brannen wrote*: Peter Brannen, *The Ends of the World: Volcanic Apocalypses, Lethal Oceans, and Our Quest to Understand Earth's Past Mass Extinctions* (New York: Ecco, 2017): 219–48.

129 *causes organic compounds to oxidize*: Chad M. Rigsby, Daniel A. Herms, Pierluigi Bonello, and Don Cipollini, "Higher Activities of Defense-Associated Enzymes May Contribute to Greater Resistance of Manchurian Ash to Emerald Ash Borer Than a Closely Related and Susceptible Congener," *Journal of Chemical Ecology* 42 (2016): 782–92.

129 *ash trees planted in China*: Liu et al., "Exploratory Survey for the Emerald Ash Borer," 192.

130 *still woodpeckers*: David E. Jennings, Juli R. Gould, John D. Vandenberg, Jian J. Duan, and Paula M. Shrewsbury, "Quantifying the Impact of Woodpecker Predation on Population Dynamics of the Emerald Ash Borer (*Agrilus planipennis*)," *PLOS One* 8, no. 6 (2013).

130 *attacking bronze birch borers*: Leah S. Bauer, Jian J. Duan, and Juli R. Gould, "Emerald Ash Borer (*Agrilus planipennis* Fairmaire) (Coleoptera: Buprestidae)," in *The Use of Classical Biological Control to Preserve Forests in North America*, ed. Roy Van Driesche and Richard Reardon, Publication FHTET-2013–2 (Morgantown, W.Va.: USDA Forest Service, 2014), 189–209.

131 *Heinrich Anton de Bary*: Anton De Bary, "The Phenomenon of Symbiosis," lecture delivered at the meeting of the German natural scientists and physicians in

Cassel (1878), in "English Translation of Heinrich Anton de Bary's 1878 speech, 'Die Erscheinung der Symbiose' ('De la symbiose')," by Nathalie Oulhen, Barbara Schulz, Tyler J. Carrier, *Symbiosis* 69 (2016): 131–39.

131 *and one or more species of fungus*: Toby Spribille, Veera Tuovinen, Philipp Resl, Dan Vanderpool, Heimo Wolinski, M. Catherine Aime, Kevin Schneider, Edith Stabentheiner, Merje Toome-Heller, Göran Thor, Helmut Mayrhofer, Hanna Johannesson, and John P. McCutcheon, "Basidiomycete Yeasts in the Cortex of Ascomycete Macrolichens," *Science* 353, no. 6298 (2016): 488–92.

131 *mycorrhizal fungi*: Paola Bonfante and Andrea Genre, "Mechanisms Underlying Beneficial Plant–Fungus Interactions in Mycorrhizal Symbiosis," *Nature Communications* 1, no. 48 (2010).

131 *We are covered in*: Rebecca A. Hall and Mairi Noverr, "Fungal Interactions with the Human Host: Exploring the Spectrum of Symbiosis," *Current Opinion in Microbiology* 40 (2017): 58–64.

131 *and filled with*: Emiley A. Eloe-Fadrosh and David A Rasko, "The Human Microbiome: From Symbiosis to Pathogenesis," *Annual Review of Medicine* 64 (2013): 145–63.

131 *"feed from the crumbs"*: de Bary, "Phenomenon of Symbiosis," 134.

131 *"beneficent and omnipotent God"*: Charles Darwin, "To Asa Gray, 22 May 1860," Gray Herbarium of Harvard University via the Darwin Correspondence Project, https://www.darwinproject.ac.uk/letter/?docId=letters/DCP-LETT-2814.xml;query=beneficent%20and%20omnipotent;brand=default.

131 *inhabit oak galls*: Richard Robinson Askew, "On the Biology of the Inhabitants of Oak Galls of Cynipidae (Hymenoptera) in Britain," *Transactions of the Society for British Entomology* 14, part 11 (1961): 237–69.

132 *a wasp of equal size*: C. E. Pemberton, "An Egg Parasite of Thrips in Hawaii," *Hawaiian Entomological Society* 7, no. 3 (1931): 481–82.

132 *"ecological fitting"*: D. H. Janzen, "On Ecological Fitting," *Oikos* 45, fasc. 3 (1985): 308–10.

133 *"The current state of the Earth system"*: John Williams and Stephen Jackson, "Novel Climates, No-Analog Communities, and Ecological Surprises," *Frontiers in Ecology and the Environment* 5, no. 6 (2007): 475–82.

134 *absolute number of species fell*: Alycia L. Stigall, "Speciation Collapse and Invasive Species Dynamics During the Late Devonian 'Mass Extinction,'" *GSA Today* 22, no. 1 (2012): 4–9.

134 *"Homogenocene"*: Michael J. Samways, "Translocating Fauna to Foreign Lands: Here Comes the Homogenocene," *Journal of Insect Conservation* 3 (1999): 65–66.

134 *"planet of weeds"*: David Quammen, "Planet of Weeds," *Harper's Magazine* (October 1998).

135 *Looking 10 million years into the future*: Chris D. Thomas, *Inheritors of the Earth: How Nature Is Thriving in an Age of Extinction* (New York: PublicAffairs, 2017).

135 *a new era of mass extinction*: Elizabeth Kolbert, *The Sixth Extinction: An Unnatural History* (New York: Henry Holt, 2014).

135 *assessing the extinction risk*: Chris D. Thomas, Alison Cameron, Rhys E. Green, Michel Bakkenes, Linda J. Beaumont, Yvonne C. Collingham, Barend F. N. Erasmus, Martinez Ferreira de Siqueira, Alan Grainger, Lee Hannah, Lesley Hughes, Brian Huntley, Albert S. Van Jaarsveld, Guy F. Midgley, Lera Miles, Miguel A. Ortega-Huera, A. Townsend Peterson, Oliver L. Phillips, and Stephen E. Williams, "Extinction Risk from Climate Change," *Nature* 427, no. 6970 (2004): 145–48.

135 *once happened to the eastern hemlock*: Margaret Bryan Davis, "Outbreaks of Forest Pathogens in Quaternary History," *Proceedings of the International Palynology Conference, Lucknow* 3 (1981): 216–28.

137 *a tree-length task*: Susan Freinkel, *American Chestnut: The Life, Death, and Rebirth of a Perfect Tree* (Berkeley: University of California Press, 2007).

138 *mine low-grade ore*: Jeanne Romero-Severson and Jennifer Koch, "Saving Green Ash," in *Proceedings of Workshop on Gene Conservation of Tree Species—Banking on the Future* (May 16–19, 2016, Chicago), 102–10.

139 *The first genetically engineered crop*: Jorge Fernandez-Cornejo, Seth Wechsler, Mike Livingston, and Lorraine Mitchell, *Genetically Engineered Crops in the United States*, Economic Research Report Number 162 (United States Department of Agriculture, 2014), 1–6.

139 *rare for forest trees*: Food and Agriculture Organization of the United Nations, *Forests and Genetically Modified Trees* (Rome: Food and Agriculture Organization of the United Nations, 2010), 10.

139 *entirely absent*: Michael A. Thomas, Gary W. Roemer, C. Josh Donlan, Brett G. Dickson, Marjorie Matocq, and Jason Malaney, "Gene Tweaking for Conservation," *Nature* 501 (September 26, 2013): 485–86.

139 *report on the conference*: Janet Marinelli, "For Endangered Florida Tree, How Far to Go to Save a Species?" *Yale Environment 360* (March 27, 2018), https://e360.yale.edu/features/for-endangered-florida-tree-how-far-to-go-to-save-a-species-torreya.

140 *"this novel approach"*: Samantha Grenrock, "Scientists Outline Novel Approach to Save Endangered Torreya Tree," *Tallahassee Democrat* (April 5, 2018), https://www.tallahassee.com/story/life/home-garden/2018/04/05/scientists-outline-novel-approach-save-endangered-torreya-tree/490402002.

141 *means to pay for itself*: Chuck Wooster, "Two Centuries of Timber and Trampers: Where Recreation and Logging Coexist," *Northern Woodlands* (Summer 2006), https://northernwoodlands.org/articles/article/two_centuries_of_timber_and_trampers_where_recreation_and_logging_coexist.

142 *three broad strategies*: C. L. Millar, N. L. Stephenson, and S. L. Stephens, "Climate Change and Forests of the Future: Managing in the Face of Uncertainty," *Ecological Applications* 17, no. 8 (2007): 2145–51.

144 *Native Americans were lighting fires*: Hazel R. Delcourt and Paul A. Delcourt, "Pre-Columbian Native American Use of Fire on Southern Appalachian Landscapes," *Conservation Biology* 11, no. 4 (1997): 1010–14.

145 *"rather like ships lost at sea"*: Deborah Rabinowitz, "Seven Forms of Rarity," *The Biological Aspects of Rare Plant Conservation*, ed. Hugh Synge (Chichester, N.Y.: John Wiley & Sons, 1981), 205–17.

6. The Future

147 *The two defining characteristics of a tree*: Colin Tudge, *The Tree: A Natural History of What Trees Are, How They Live, and Why They Matter* (New York: Three Rivers, 2005), 65.

147 *"aesthetes and engineers"*: Ibid., 75.

148 *"every age has been a Wood Age"*: Ibid., xvi.

148 *People have long worried*: Michael Williams, *Deforesting the Earth: From Prehistory to Global Crisis* (Chicago: University of Chicago Press, 2003), 170.

148 *"a public calamity"*: John Evelyn, *Silva; or, A Discourse of Forest-Trees*, vol. 2 (1662; repr., York, England: J. Todd, 1786), 262.

148 *"to possesse but 50 Acres"*: William Cronon, *Changes in the Land: Indians, Colonists, and the Ecology of New England* (1983; repr., New York: Hill and Wang, 2003), 25.

148 *"unfit home for its noblest inhabitant"*: George Perkins Marsh, *The Earth as Modified by Human Action: A Last Revision of "Man and Nature"* (1864; repr., New York: Charles Scribner's Sons, 1898), 43.

148 *a report by William Hall*: William L. Hall, *The Waning Hardwood Supply and the*

Appalachian Forests, U.S. Department of Agriculture Forest Service-Circular 116 (Washington, D.C.: Government Printing Office, 1907).

149 *due to inertia*: Kees van der Heijden, "Scenarios, Strategies and the Strategy Process," Nijenrode Research Paper Series Centre for Organisational Learning and Change no. 1997–01 (Breukelen, Netherlands: Nijenrode University Press, 1997), http://www.nijenrode.nl/library/publications/nrp/1997–01/1997–01.html.

150 *"no great commercial disaster"*: A. S. Mather, "Global Trends in Forest Resources," *Geography* 72, no. 1 (January 1987): 1–15.

150 *by roughly 2.7 degrees*: Intergovernmental Panel on Climate Change, *Global Warming of 1.5 Degrees C: Summary for Policymakers* (Switzerland: Intergovernmental Panel on Climate Change, 2018).

150 *higher still*: Intergovernmental Panel on Climate Change, *Climate Change 2014 Synthesis Report: Summary for Policy Makers* (Switzerland: Intergovernmental Panel on Climate Change, 2014), 12.

151 *"terrifying eeriness, awe, and horror"*: Williams, *Deforesting the Earth*, 163.

152 *more purposeful harvest*: Elizabeth May, *At the Cutting Edge: The Crisis in Canada's Forests* (Toronto: Key Porter Books, 1998), 114.

152 *"steady expansion"*: Ken Drushka, *Canada's Forests: A History* (Montreal: McGill–Queen's University Press, 2003), 63.

155 *the image of Quebec*: Gouvernement du Québec, "Forêt Ouverte" (Gouvernement du Québec, 2019), https://www.foretouverte.gouv.qc.ca.

155 *"In that Empire"*: Jorge Luis Borges, "On Exactitude in Science," in *Collected Fictions*, trans. Andrew Huxley (New York: Penguin Books, 1998), 325.

156 *some four hundred thousand plots*: Louis Duchesne, private communications.

157 *oldest comprehensive satellite images*: Alan S. Belward and Jon O. Skøien, "Who Launched What, When and Why: Trends in Global Land-Cover Observation Capacity from Civilian Earth Observation Satellites," *ISPRS Journal of Photogrammetry and Remote Sensing* 103 (2013): 115–28.

157 *building one carbon-storage map*: Patrick Gonzalez, John J. Battles, Brandon M. Collins, Timothy Robards, and David S. Saah, "Aboveground Live Carbon Stock Changes of California Wildland Ecosystems, 2001–2010," *Forest Ecology and Management* 348 (2015): 68–77.

158 *trees dying at increasing rates*: Phillip J. van Mantgem, Nathan L. Stephenson, John C. Byrne, Lori D. Daniels, Jerry F. Franklin, Peter Z. Fulé, Mark E. Harmon, Andrew J. Larson, Jeremy M. Smith, Alan H. Taylor, and Thomas T. Veblen, "Widespread

Increase of Tree Mortality Rates in the Western United States," *Science* 323 (2009): 521–24.

158 *uprooted small piñon pines*: Henry D. Adams, Maite Guardiola-Claramonte, Greg A. Barron-Gafford, Juan Camilo Villegas, David D. Breshears, Chris B. Zou, Peter A. Troch, and Travis E. Huxman, "Temperature Sensitivity of Drought-Induced Tree Mortality Portends Increased Regional Die-off Under Global-Change-Type Drought," *Proceedings of the National Academy of Sciences* 106, no. 17 (2009): 7063–66.

159 *the insects that attack trees*: Adrian J. Das, Nathan L. Stephenson, and Kristen P. Davis, "Why Do Trees Die? Characterizing the Drivers of Background Mortality," *Ecology* 97, no. 10 (2016): 2616–27.

159 *spreading poleward and upslope*: Corey Lesk, Ethan Coffel, Anthony W. D'Amato, Kevin Dodds, and Radley Horton, "Threats to North American Forests from Southern Pine Beetle with Warming Winters," *Nature Climate Change* 7 (2017): 713–18.

159 *new species of trees*: L. Safranyik, A. L. Carroll, J. Régnière, D. W. Langor, W. G. Riel, T. L. Shore, B. Peter, B. J. Cooke, V. G. Nealis, and S. W. Taylor, "Potential for Range Expansion of Mountain Pine Beetle into the Boreal Forest of North America," 142, no. 5 (2010): 415–42.

159 *the grip of a budworm outbreak*: Yan Boulanger and Dominique Arsenault, "Spruce Budworm Outbreaks in Eastern Quebec over the Last 450 Years," *Canadian Journal of Forest Research* 34, no. 5 (2004): 1035–43.

160 *roughly 4 percent of the world's land burns*: Stefan H. Doerr and Cristina Santín, "Global Trends in Wildfire and Its Impacts: Perceptions Versus Realities in a Changing World," *Philosophical Transactions of the Royal Society B* 371 (2016).

160 *the fire season*: W. Matt Jolly, Mark A. Cochrane, Patrick H. Freeborn, Zachary A. Holden, Timothy J. Brown, Grant J. Williamson, and David M. J. S. Bowman, "Climate-Induced Variations in Global Wildfire Danger from 1979 to 2013," *Nature Communications* 6, no. 7537 (2015).

162 *a rock hit Earth*: Martin Schmieder, Barry J. Shaulis, Thomas J. Lapen, Elmar Buchner, and David A. Kring, "In Situ U–Pb Analysis of Shocked Zircon from the Charlevoix Impact Structure, Québec, Canada," *Meteoritics & Planetary Science* 54, no. 8 (2019): 1808–27.

163 *hundreds of miles out of place*: J. P. Paul Jasinski and Serge Payette, "The Creation of Alternative Stable States in the Southern Boreal Forest, Quebec, Canada," *Ecological Monographs* 75, no. 4 (2005): 561–83.

163 *arranged the way they are*: Jorge Soberón and A. Townsend Peterson, "Interpretation of Models of Fundamental Ecological Niches and Species' Distributional Areas," *Biodiversity Informatics* 2 (2005): 1–10.

163 *"a large shift in some systems"*: Marten Scheffer and Stephen R. Carpenter, "Catastrophic Regime Shifts in Ecosystems: Linking Theory to Observation," *Trends in Ecology and Evolution* 18, no. 12 (2003): 648–56.

164 *one fire, followed a few decades later*: C. D. Brown and J. F. Johnstone, "Once Burned, Twice Shy: Repeat Fires Reduce Seed Availability and Alter Substrate Constraints on Black Spruce Regeneration," *Forest Ecology and Management* 266 (2012): 34–41.

164 *spruce budworm, then fire*: Jasinski and Payette, "Creation of Alternative Stable States."

164 *had turned into lichen woodland*: F. Girard, S. Payette, and R. Gagnon, "Rapid Expansion of Lichen Woodlands Within the Closed-Crown Boreal Forest Zone over the Last 50 Years Caused by Stand Disturbances in Eastern Canada," *Journal of Biogeography* 35 (2008): 529–37.

164 *fall from dominance*: Zelalem A. Mekonnen, William J. Riley, James T. Randerson, Robert F. Grant, and Brendan M. Rogers, "Expansion of High-Latitude Deciduous Forests Driven by Interactions Between Climate Warming and Fire," *Nature Plants* 5, no. 9 (2019).

165 *transform with similar speed*: Constance I. Millar and Nathan L. Stephenson, "Temperate Forest Health in an Era of Emerging Megadisturbance," *Science* 349, no. 6250 (2015): 823–26.

166 *yews grown in Sweden*: David Boshier, Linda Broadhurst, Jonathan Cornelius, Leonardo Gallo, Jarkko Koskela, Judy Loo, Gillian Petrokofsky, and Bradley St. Clair, "Is Local Best? Examining the Evidence for Local Adaptation in Trees and Its Scale," *Environmental Evidence* 4, no. 20 (2015): 1–10.

167 *"If global warming materializes as projected"*: F. Thomas Ledig and J. H. Kitzmiller, "Genetic Strategies for Reforestation in the Face of Global Climate Change," *Forest Ecology and Management* 50 (1992): 153–69.

167 *British Columbia, Alberta, and Quebec*: Adam Wellstead and Michael Howlett, "Assisted Tree Migration in North America: Policy Legacies, Enhanced Forest Policy Integration, and Climate Change Adaptation," *Scandinavian Journal of Forest Research* 32, no. 6 (2017): 535–43.

167 *Forest Inventory and Analysis Plots*: U.S. Forest Service, *Forest Inventory and Analysis Fiscal Year 2017 Business Report* (USDA, Forest Service, May 2019), 5.

168 *to build a model*: F. Thomas Ledig, Gerald E. Rehfeldt, and Barry Jaquish, "Projections of Suitable Habitat Under Climate Change Scenarios: Implications for Trans-Boundary Assisted Migration," *American Journal of Botany* 99, no. 7 (2012): 1–14.

168 *just fade and disappear*: Nick Crookston, "Plant Species and Climate Profile Predictions," Nick Crookston's personal website; it migrated from the Forest Service server after Crookston's retirement: http://charcoal.cnre.vt.edu/climate.

169 *"the species' genetic diversity"*: Ledig, Rehfeldt, and Jaquish, "Projections of Suitable Habitat," 11.

169 *a series of common gardens*: Guillaume Otis Prud'homme, Mohammed S. Lamhamedi, Lahcen Benomar, André Rainville, Josianne DeBlois, Jean Bousquet, and Jean Beaulieu, "Ecophysiology and Growth of White Spruce Seedlings from Various Seed Sources Along a Climatic Gradient Support the Need for Assisted Migration," *Frontiers in Plant Science* 8, art. 2214 (2018).

170 *a series of chemical stomachs*: Pratima Bajpai, "Brief Description of the Pulp and Papermaking Process," in *Biotechnology for Pulp and Paper Processing*, 2nd ed. (Singapore: Springer Nature Singapore, 2018), 9–26.

171 *Different people and organizations*: Jasmin Guénette and Pierre Desrochers, "Are Quebec's Forests Threatened?" *Economic Note* (Montreal Economic Institute, 2014), http://www.iedm.org/sites/default/files/pub_files/note0714_en.pdf.

171 *"forest transition"*: A. S. Mather, "The Forest Transition," *Area* 24, no. 4 (1992): 367–79.

171 *roughly half the world's trees*: T. W. Crowther, H. B. Glick, K. R. Covey, C. Bettigole, D. S. Maynard, S. M. Thomas, J. R. Smith, G. Hintler, M. C. Duguid, G. Amatulli, M.-N. Tuanmu, W. Jetz, C. Salas, C. Stam, D. Piotto, R. Tavani, S. Green, G. Bruce, S. J. Williams, S. K. Wiser, M. O. Huber, G. M. Hengeveld, G.-J. Nabuurs, E. Tikhonova, P. Borchardt, C.-F. Li, L. W. Powrie, M. Fischer, A. Hemp, J. Homeier, P. Cho, A. C. Vibrans, P. M. Umunay, S. L. Piao, C. W. Rowe, M. S. Ashton, P. R. Crane, and M. A. Bradford, "Mapping Tree Density at a Global Scale," *Nature* 525 (2015): 201–5.

172 *Manic 5 began to fill*: Lili Réthi and William W. Jacobus Jr., *Manic 5: The Building of the Daniel Johnson Dam* (New York: Doubleday, 1971).

172 *the meteorite was about five miles across*: P. I. K. Onorato, D. R. Uhlmann, and C. H. Simonds, "The Thermal History of the Manicouagan Impact Melt Sheet, Quebec," *Journal of Geophysical Research* 83, no. B6 (1978): 2789–98.

173 *When the meteorite hit*: Evelyn Kustatscher, Sidney Ash, Eugeny Karasev, Christian Pott, Vivi Vajda, Jianxin Yu, and Stephen McLoughlin, "Flora of the Late Triassic,"

in *The Late Triassic World: Earth in a Time of Transition*, ed. Tanner Lawrence (New York: Springer, 2018), 522–622.

7. Where to Plant a Tree

175 *journalist Jim Robbins*: Jim Robbins, "Tall, Ancient and Under Pressure," *New York Times* (August 11, 2014), https://www.nytimes.com/2014/08/12/science/tall-ancient-and-under-pressure.html.

177 *Alexander Smith*: Alexander Smith, *Dreamthorp: A Book of Essays Written in the Country* (London: Strahan & Co., 1863), 258.

177 *John Boyle*: John Boyle, *The Orrery Papers*, vol. 2, ed. Countess of Cork and Orrery (London: Duckworth, 1903): 56.

177 *Susan Fenimore Cooper*: Susan Fenimore Cooper, *Rural Hours* (New York: George P. Putnam, 1850), 217.

177 *John Evelyn*: John Evelyn, *Silva; or, A Discourse of Forest Trees*, vol. 2 (1662; repr., York, England: J. Todd, 1786), 340.

179 *put together the plan*: Sierra Pacific Industries, "Giant Sequoia Genetic Conservation Plan Progress Report—Sierra Pacific Industries, Updated July, 2015," http://www.torreyaguardians.org/sequoia-lunak-2015.pdf.

180 *"last great lumber baron"*: Chloe Sorvino, "A Billion-Dollar Fortune Made from Timber and Fire," *Forbes* (May 14, 2018), https://www.forbes.com/feature/archie-emmerson-timber-forest-fires-logging/#7ed61c1364f9.

180 *mechanical properties similar to those*: Douglas D. Piirto, "Wood of Giant Sequoia: Properties and Unique Characteristics," in *Proceedings of the Workshop on Management of Giant Sequoia*, May 24–25, 1985, Reedley, California (Berkeley: U.S. Forest Service, Pacific Southwest Forest and Range Experiment Station, 1986), 19–23.

181 *"The true meaning of life"*: Wes Henderson, *Under Whose Shade: A Story of a Pioneer in the Swan River Valley of Manitoba* (Nepean, Ont.: Wes Henderson & Associates, 1986).

184 *"climate awakening"*: Christopher Mims and Stephanie Gruner Buckley, "5 Charts About Climate Change That Should Have You Very, Very Worried," *Atlantic* (November 24, 2012), https://www.theatlantic.com/technology/archive/2012/11/5-charts-about-climate-change-that-should-have-you-very-very-worried/265554.

185 *had compiled a list*: Paul Camire, "Florida Torreya: Ex Situ Specimens of *Torreya taxifolia*" (September 18, 2018), http://www.torreyaguardians.org/historic-list.pdf.

186 *petition to downlist*: Connie Barlow, "Petition to DOWNLIST from Endangered to Threatened *Torreya taxifolia*—Florida Torreya" (September 9, 2019).

186 *More than a hundred papers*: N. Hewitt, N. Klenk, A. L. Smith, D. R. Bazely, N. Yan, S. Wood, J. I. MacLellan, C. Lipsig-Mumme, and I. Henriques, "Taking Stock of the Assisted Migration Review," *Biological Conservation* 144 (2011): 2560–72.

186 *tuatara*: Kimberly A. Miller, Hilary C. Miller, Jennifer A. Moore, Nicola J. Mitchell, Alison Cree, Fred A. Allendorf, Stephen D. Sarre, Susan N. Keall, and Nocola J. Nelson, "Securing the Demographic and Genetic Future of Tuatara Through Assisted Colonization," *Conservation Biology* 26, no. 5 (2012): 790–98.

186 *Farsi tooth-carp*: Hamid Reza Esmaeli, Mojtaba Masoudi, Mehregan Ebrahimi, and Amir Elmi, "Review of *Aphanius farsicus:* A Critically Endangered Species (Teleostei: Cyprinodontidae) in Iran," *Iranian Journal of Ichthyology* 3, no. 1 (2016): 1–18.

186 *American pika*: Jennifer L. Wilkening, Chris Ray, Nathan Ramsay, and Kelly Klingler, "Alpine Biodiversity and Assisted Migration: The Case of the American Pika (*Ochotona princeps*)," *Biodiversity* 16, no. 4 (2015): 224–36.

187 *a butterfly here*: Stephen G. Willis, Jane K. Hill, Chris D. Thomas, David B. Roy, Richard Fox, David S. Blakeley, and Brian Huntley, "Assisted Colonization in a Changing Climate: A Test-Study Using Two U.K. Butterflies," *Conservation Letters* 2, no. 1 (2009): 46–52.

187 *a common garden there*: S. C. McLane and S. N. Aitken, "Whitebark Pine (*Pinus albicaulis*) Assisted Migration Potential: Testing Establishment North of the Species Range," *Ecological Applications* 22, no. 1 (2012): 142–53.

187 *the latest draft*: A. Karasov-Olson, M. W. Schwartz, S. Skikne, J. D. Olden, J. Hellmann, C. H. Hoffman, G. Schuurman, P. Gonzalez, D. J. Lawrence, S. Allen, M. Trammell, D. Buttke, A. Miller-Rushing, C. Brigham, and J. Morisette, "NPS Ecological Risk Assessment of Managed Relocation as a Climate Change Adaptation Strategy Workbook" (draft, private communications with Mark Schwartz, 2019).

189 *a man planting a carob tree*: Michael L. Rodkinson, *New Edition of the Babylonian Talmud*, vol. 8 (New York: New Talmud Publishing, 1899), 65–66.

190 *"the air in a close apartment*: Marsh, *Man and Nature*, 153.

190 *the Green Belt Movement*: Norwegian Nobel Institute, "Wangari Maathai Biographical," 2004, https://www.nobelprize.org/prizes/peace/2004/maathai/biographical.

191 *fifty thousand trees*: Henry de Quetteville, "The 13-Year-Old Who Has the World Planting Trees," *Telegraph* (April 29, 2011), https://www.telegraph.co.uk/news/earth/8476747/The-13-year-old-who-has-the-world-planting-trees.html.

191 *Paris Agreement*: United Nations, *Paris Agreement* (Paris, 2015), 6.

191 *13.6 billion trees*: Trillion Tree Campaign, trilliontreecampaign.org.

191 *other ways of slowing climate change*: Royal Society, *Geoengineering the Climate: Science, Governance, and Uncertainty* (London: Royal Society, September 2009), x.

191 *practical and innocuous*: Rob Bellamy, Jason Chilvers, Naomi E. Vaughan, and Timothy M. Lenton, "A Review of Climate Geoengineering Appraisals," *Wiley Interdisciplinary Reviews: Climate Change* 3, no. 6 (2012): 597–615.

191 *Bonn Challenge*: Simon L. Lewis, Charlotte E. Wheeler, Edward T. A. Mitchard, and Alexander Koch, "Regenerate Natural Forests to Store Carbon," *Nature* 568 (2019): 25–28.

192 *0.9 billion hectares*: Jean-Francois Bastin, Yelena Finegold, Claude Garcia, Danilo Mollicone, Marcelo Rezende, Devin Routh, Constantin M. Zohner, and Thomas W. Crowther, "The Global Tree Restoration Potential," *Science* 365, no. 6448 (2019): 76–79.

194 *a letter to President Franklin Roosevelt*: Albert Einstein, "Letter to President F. D. Roosevelt" (via Franklin D. Roosevelt Presidential Library and Museum, August 2, 1939), http://www.fdrlibrary.marist.edu/archives/pdfs/docsworldwar.pdf.

INDEX